Fundamentals of Optical Fiber Communications

SECOND EDITION

Academic Press Rapid Manuscript Reproduction

Fundamentals of Optical Fiber Communications

SECOND EDITION

Edited by

Michael K. Barnoski
TRW Technology Research Center
El Segundo, California

ACADEMIC PRESS

A Subsidiary of Harcourt Brace Jovanovich, Publishers

New York London Toronto Sydney San Francisco 1981

ACADEMIC PRESS, INC.
111 Fifth Avenue, New York, New York 10003

United Kingdom Edition published by
ACADEMIC PRESS, INC. (LONDON) LTD.
24/28 Oval Road, London NW1   7DX

Library of Congress Cataloging in Publication Data
Main entry under title:

Fundamentals of optical fiber communications.   .

    Contents: Optical fiber waveguides / Donald B. Keck --
Optical fiber cabel / James E. Goell -- Coupling components
for optical fiber waveguides / Michael K. Barnoski -- [etc.]
    1. Optical communications. 2. Fiber optics.
I. Barnoski, Michael K., Date.
TK5103.59.F86 1981   621.38'414   81-12883
ISBN 0-12-079151-X   AACR2

Contents

7 DESIGN CONSIDERATIONS FOR MULTITERMINAL NETWORKS

M. K. Barnoski

Contributors

Numbers in parentheses indicate the pages on which authors' contributions begin.

MICHAEL K. BARNOSKI (147, 329), TRW Technology Research Center, El Segundo, California 90245

JAMES E. GOELL (109) Lightwave Technologies, Inc., Van Nuys, California 91406

DONALD B. KECK (1), Corning Glass Works, Corning, New York 14831

HENRY KRESSEL (187), RCA Laboratories, Princeton, New Jersey 08540

STEWART D. PERSONICK (257, 295), TRW Technology Research Center, El Segundo, California 90245

Preface

The achievement of low loss transmission has made the optical fiber waveguide the leading contender as the transmission medium for a vast variety of current and future systems. Since the publication of the first edition of this text, the technology has advanced from research and development and the early system trial stage to the point where, because it is the most appropriate technology, it is now being incorporated into systems.

The material presented in this book is intended to provide the reader with a tutorial treatment of fiber optic technology as applied to communications systems. The text is based on lectures presented at an annual short course entitled "Fiber Optic Communication Systems" at the University of California at Santa Barbara. This course, which has been presented every summer for the past seven years, provides an opportunity for the incorporation of student recommendations into both the content of material and the method of presentation.

The editor gratefully acknowledges the contributing authors and the institutions with which they are associated for their wholehearted cooperation in the preparation of this book. The most competent assistance of Mrs. Shirley Biamonte, Mrs. Ann Garcia, Mrs. Nancy Fox, and Mr. Seichi Kiyohara in the preparation of this text is very much appreciated.

M. K. Barnoski

CHAPTER 1

OPTICAL FIBER WAVEGUIDES

Donald B. Keck

Research and Development Laboratory
Corning Glass Works
Corning, New York   14831

1.1   INTRODUCTION

The achievement of low-loss transmission coupled with other attendent advantages, such as large information carrying capacity, immunity from electromagnetic interference, dielectric conductor, small size and weight, has literally created a new technology in which the optical fiber waveguide will be increasingly used as the transmission medium for a wide variety of applications. Although the concept of light transmission over a dielectric conductor dates back to John Tyndell (1.1), who in 1854, performed experiments before the Royal Society, and the possibility of telecommunications using light waves was posed by Alexander Graham Bell in 1880 (1.2), it has only been since about 1970 that the necessary components have been available to seriously consider optical fiber systems. In the telecommunications area, these range from comparatively short length, a few hundred meter applications, to submarine installations in which repeater spacings of 50 to 100 km are expected. Additionally, very sophisticated sensor applications based on preservation of phase and polarization information over long distances are emerging. The environmental aspect of the optical fiber waveguides and associated cables and buffers, such as strength, fatigue, durability, and thermal behavior, have received much attention with positive results.

Whereas telecommunications is expected to be the dominant application, the coherent propagation possible in the optical fiber waveguide is a driving force for the entire field of integrated optics. The technology spawned by these low-attenuation, high-information capacity optical conductors holds the promise of revolutionizing our way of life by allowing wide-spread access to large quantities of information and allowing heretofore unreachable levels of interactive communication. The future for optical fiber communications is extremely promising. In the ensuing pages, the key elements of this technology will be discussed in considerable detail by several of the pioneers in this field.

Many types of waveguiding structures have been proposed in the literature, such as the single material (1.3) and W-type (1.4) guides, each with its own advantages and disadvantages. Throughout, however, the most universally applicable waveguide type is the single high-refractive index solid core surrounded by a lower refractive index solid cladding. It is this structure that will be examined, with particular emphasis on its long-distance propagation characteristics.

This chapter is divided into four main sections. In the first section the ideal waveguide will be discussed. This will begin with a ray description to obtain an intuitive picture of the propagation, move on to the detailed solution of Maxwell's equations for the cylindrical waveguide, and end with the WKBJ solution which furnishes a powerful approximate solution. The correspondence between a ray and modal description of the waveguide is presented utilizing a phase space model.

The second section brings in a degree of imperfection to the waveguide in the form of the constituent materials where the various sources of waveguide attenuation are discussed.

Information carrying capacity in the ideal waveguide
is the topic of the third section with the effect of the
waveguide materials being included in the analysis. This discussion makes use of the results of the WKBJ solution from the
first section. The final section deals with propagation in a
perturbed waveguide which can lead to intermodal coupling of
power. The perturbations may be intrinsically part of the
waveguide or result unintentionally from attempts to package
or deploy the waveguide structure. The resulting mode coupling potentially leads to an attenuation-bandwidth system
tradeoff, which is discussed.

1.2   PROPAGATION IN IDEAL FIBER WAVEGUIDES

1.2.1   Ray Description
Since the early 1900's physics has been faced with
the duality of rays and waves. For most problems involving
electromagnetic propagation one has found that ray formalism,
while not incorrect, was not best suited for explaining the
details of the physical phenomenon involved. This is also the
case for the cylindrical fiber waveguide as we shall see. Ray
optics, however, does provide a simpler picture for describing
waveguide operation, and therefore warrants discussion.

1.2.1.1   Homogeneous Core (Step) Fiber
Let us consider first the simple step index waveguide
which, as shown in axial cross section in Figure 1.1, consists
of a uniform core region of diameter 2a and index, n, surrounded
by a cladding of index, $n(1-\Delta)$, where Δ is approximately the
fractional index difference between core and cladding.

Two types of rays exist in such a structure; meridional
rays, which pass through the guide axis, and skew rays, which
do not. A typical meridional ray is shown in the figure. As
long as the external angle, θ_o, that it makes with the guide

axis is less than θ_c where,

$$\sin\theta_o \leq \sin\theta_c \cong \frac{n\sqrt{2\Delta}}{n_e} \, , \qquad (1.1)$$

the ray will be totally internally reflected at the core/cladding
interface. Here terms of order Δ^2 have been neglected, which is
a valid assumption since for many telecommunications waveguides,
$\Delta = 0.01$.

In the traditional optical sense, the numerical
aperture of the waveguide is defined,

$$NA = \sin\theta_c = \frac{n\sqrt{2\Delta}}{n_e} \, , \qquad (1.2)$$

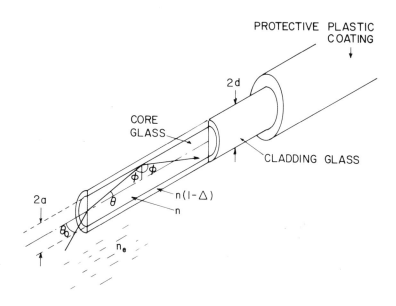

Figure 1.1.   Typical meridional ray in a step refractive index
             fiber.

assuming a waveguide in air (n_e = 1) with n = 1.46 and
Δ = 0.01, the maximum angle ray accepted by the waveguide is

$$\theta_c = \sin^{-1} [1.46 \ (2 \times 0.01)^{1/2}] \cong 12 \text{ degrees} . \quad (1.3)$$

This is a typical value for waveguides used for telecommunications applications. With simple geometry the path
length, ℓ, of a meridional ray, can also be obtained:

$$\ell(\theta) = L \sec \theta \ , \quad\quad\quad (1.4)$$

where, L is the axial length of the guide.

The path length, and therefore the transit time, is
a function of the ray angle. This differential delay between
the modes reduces the information capacity of the waveguide.
It also renders the step index guide incapable of imaging.

The general skew ray is shown in Figure 1.2. It is
incident on the end of the guide at a point $\overrightarrow{r_0}$ = x_0 $\hat{i}$ + y_0 $\hat{j}$ =
$|\overrightarrow{r_0}|$ [cos ψ_0 $\hat{i}$ + sin ψ_0 $\hat{j}$] where,$|\overrightarrow{r_0}|$= $(x_0{}^2 + y_0{}^2)^{\frac{1}{2}}$, and at
an angle defined by the unit vector, $\hat{q}_0$ = $\overrightarrow{Q_0}/|\overrightarrow{Q_0}|$= $\sin\theta_0$ $\cos\phi_0$ $\hat{i}$
+ $\sin\theta_0$ $\sin\phi_0$ $\hat{j}$ + $\cos\theta_0$ $\hat{k}$. The elemental spatial and angular
areas dA and dΩ through which the ray passes will be needed in
subsequent discussion of the radiation field intensities. If $\hat{q}_m$
and $\overrightarrow{r}_m$ describe the ray vector prior to the mth reflection at the
core/cladding interface, then the vector relations

$$(\hat{q}_m - \hat{q}_{m+1}) \times \overrightarrow{r}_m = \overrightarrow{0} \quad\quad\quad (1.5)$$

and

$$(\hat{q}_m + \hat{q}_{m+1}) \cdot \overrightarrow{r}_m = 0 \quad\quad\quad (1.6)$$

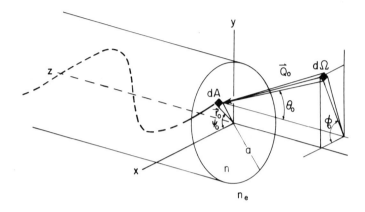

Figure 1.2.   Ray coordinate system for the cylindrical fiber.
               A ray is characterized by its angles θ and ϕ and
               cylindrical coordinates r and ψ.

must hold. The first of these is the condition for coplanarity
of the incident and reflected ray; the second is the condition
for equal angles of incidence and reflection. For total internal
reflection there is also the condition,

$$\hat{q}_m \cdot \frac{\vec{r}_m}{|\vec{r}_m|} \leq \sqrt{2\Delta} \ . \tag{1.7}$$

From these vector equations and with much algebra,
the general expression for the capture of a ray can be derived:

$$\sin^2\theta_o \left[1 - \left(\frac{r_o}{a}\right)^2 \sin^2 (\phi_o - \psi_o) \right] \leq \frac{n^2(2\Delta)}{n_e^2} \ . \tag{1.8}$$

Without loss of generality, we may consider the x axis to lie along r so that $\psi_o = 0$. In this case, the incident ray is meridional, and for the largest value of θ_o, Eq. (1.2) results. Next consider the situation $r_o = a$ in which case $\sin\theta_o \cos\phi_o \leq n \sqrt{2\Delta}/n_e$. This inequality may be satisfied even for $\theta_o \to \pi/2$ if, correspondingly, $\phi_o \to \pi/2$. That is, a ray can enter the waveguide at 90° with respect to both the x and z axis and still be confined, however, with zero axial velocity. Skew rays can indeed exist at very high angles in the waveguide. More will be said about these rays when leaky modes are considered in a later section.

1.2.1.2   Inhomogeneous Core (Graded) Fiber

Attention is next turned to the more general refractive index core and a discussion of rays in such a medium. The general ray path described by the vector $\overrightarrow{ds}$ in Figure 1.3, is found to be perpendicular to planes of constant phase, $S(x,y,z)$, where

$$(\overrightarrow{\nabla S})^2 = n^2 .$$ (1.9)

This is the Eikonal equation which is derived from Maxwell's equations. If ds is an elemental path length along the ray, then it can be shown that

$$\frac{d}{ds} \left(n \frac{\overrightarrow{dR}}{ds} \right) = \overrightarrow{\nabla n} .$$ (1.10)

Further, if the ray makes a small angle with respect to the z axis the replacement, ds → dz, can often be made, which corresponds to the paraxial approximation.

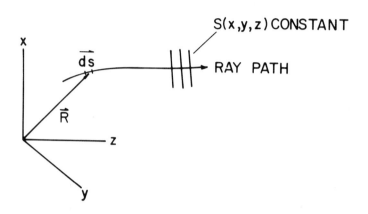

Figure 1.3.   Coordinate system for describing a ray in an
inhomogeneous medium.

To match the waveguide geometry, this equation is
cast into cylindrical coordinates. The dependence of the
unit vectors on the cylindrical coordinate variables must be
taken into account during the differentiation. With care,
the following three component equations are obtained:

$$\frac{d}{ds}\left(n\frac{dr}{ds}\right) - nr\left(\frac{d\psi}{ds}\right)^2 = \frac{dn}{dr} \ (r \text{ component})$$ (1.11a)

$$n\left(\frac{dr}{ds}\right)\left(\frac{d\psi}{ds}\right) + \frac{d}{ds}\left(nr\frac{d\psi}{ds}\right) = 0 \ (\psi \text{ component})$$ (1.11b)

$$\frac{d}{ds}\left(n\frac{dz}{ds}\right) = 0 \ (z \text{ component}) \quad .$$ (1.11c)

It has been assumed in writing these that the core refractive index, n, is a function only of the radial coordinate. The z-component can be directly integrated to give,

$$ds = \left(\frac{n}{n_o \cos\theta_o} \right) dz \quad . \tag{1.12}$$

Here, θ_o and r_o are the initial angle and radial position of the ray, and $n_o = n(r_o)$. Using this result the ψ component can be rewritten

$$r^2 \frac{d\psi}{dz} = \frac{1}{\cos\theta_o} (x_o \sin\theta_o \sin\phi_o - y_o \sin\theta_o \cos\phi_o) \tag{1.13}$$

Finally substituting both of these into the r component and integrating once we obtain

$$\tag{1.14}$$

$$z = \int_{r_o}^{r} \frac{\cos\theta_o \, dr}{\left\{ \left[\frac{n(r)}{n_o} \right]^2 + \left[1 - \frac{r_o}{r} \right]^2 (x_o \sin\theta_o \sin\phi_o - y_o \sin\theta_o \cos\phi_o) - \cos^2\theta_o \right\}^{1/2}} \quad .$$

Thus the ray path can be uniquely specified once the index distribution, $n(r)$, and the initial ray parameters, x_o, y_o, θ_o, and ϕ_o are known.

To obtain a feel for rays in a graded index medium we consider a few specific examples. Consider first the case of meridional rays. Without loss of generality we may pick $y_o = \phi_o = 0$ and the $x_o = r_o$. Eq. (1.14) then becomes

$$z = \int_{r_o}^{r} \frac{\cos\theta_o \, dr}{\left\{ \left[\frac{n(r)}{n_o} \right]^2 - \cos^2\theta_o \right\}^{1/2}} \quad . \tag{1.15}$$

Consider first the distribution $n(r)=n(0)[1-2\Delta(r/a)^2]^{\frac{1}{2}}$, the so called square law medium. Substituting this into Eq.(1.15) and evaluating the elementary integral gives the radial coordinate of the ray as a function of position

$$r = C \sin\left[\frac{\sqrt{2\Delta}}{n_o \cos\theta_o} \left(\frac{z}{a}\right)\right] \quad , \qquad (1.16)$$

where,

$$C = \frac{an_o \cos\theta_o}{\sqrt{2\Delta}} \left\{\frac{1}{\cos^2\theta_o\left[1 - 2\Delta\left(\frac{r_o}{a}\right)^2\right]} - 1\right\}^{1/2} . \qquad (1.17)$$

The ray path is periodic in z with a period,

$$\Lambda = \frac{2\pi a \cos\theta_o}{\sqrt{2\Delta}} \left[1 - 2\Delta\left(\frac{r_o}{a}\right)^2\right]^{1/2} . \qquad (1.18)$$

Thus the period depends upon both the input position, r_o, and input angle, θ_o, and is therefore different for every meridional ray. In general, as long as the index exhibits a monotonic decrease with radius, a sinusoidal path within the waveguide will result with its period determined by the initial conditions and the exact nature of the profile.

Next consider the distribution $n(r) = n(0)\text{sech}[2\Delta(r/a)]$. Again, the integral in Eq. (1.15) is of elementary form and the result is

$$\sinh\left[2\Delta\left(\frac{r}{a}\right)\right] = C'\sin\left[2\Delta\left(\frac{z}{a}\right)\right] + \sinh\left[2\Delta\left(\frac{r_o}{a}\right)\right] , \qquad (1.19)$$

where

$$C' = \left(\frac{\cosh^2\left[2\Delta\left(\frac{r_o}{a}\right)\right]}{\cos\theta_o} - 1\right)^{1/2} . \qquad (1.20)$$

Once again the path is periodic in z. Now, however, the period is independent of the initial position or angle of the ray,

$$\Lambda = \frac{\pi a}{\sqrt{8\Delta}} . \qquad (1.21)$$

This distribution has zero differential delay for all meridional rays and forms the basis for the "self-focusing" waveguides (1.5).

Thus far, only meridional rays have been considered. There is one index distribution for which the path for a special class of skew rays can be obtained exactly (1.6). The rays are helical, with the condition that $dr/dz = 0$, and the index distribution is

$$n(r) = n(o) \left[1 + 2\Delta \left(\frac{r}{a} \right)^2 \right]^{-1/2} . \qquad (1.22)$$

It is then found that the path of the ray is described by

$$\psi = \left[\frac{\sin\theta_o \ \sin\phi_o}{r_o \ \cos\theta_o} \right] z + \psi_o . \qquad (1.23)$$

Again, for this ray class we see that the path is periodic in z with a length

$$\Lambda = \frac{2\pi r_o \ \cos\theta_o}{\sin\theta_o \ \sin\phi_o} , \qquad (1.24)$$

but that it again depends on the initial position and angle of the ray. Thus dispersion will again exist between various rays of this class.

In presenting these examples, we have obtained a general picture of ray propagation in the general graded guide. It is to be noted that no single distribution is capable of simultaneously focusing all rays, and that, therefore, differential delays will always exist. It should also be noted that if $\sqrt{2\Delta}/a$ is small, these three index distribution functions are the same to first order in Δ,

$$n(r) = n(o) \left[1 - 2\Delta \left(\frac{r}{a} \right)^2 \right]^{1/2} \qquad (1.25)$$

This indicates that for small Δ, a near focus condition can be obtained resulting in small differential delays for all rays and, therefore, high information carrying capacity. In the section on information carrying capacity, we will see that a slight modification of this profile results in low dispersion.

The preceding discussion has indicated that the radially graded medium can be considered a concatenation of lenses periodically focusing the propagating rays. The imaging properties of these structures is an important subject in itself (1.6 - 1.10) and is receiving renewed interest.

1.2.2   Mode Description

1.2.2.1   General Theory

More detailed knowledge of propagation characteristics of the optical fiber waveguide can only be obtained by solution of Maxwell's equations. This leads to only certain allowed modes which can propagate in a particular dielectric structure. If the structure is such that a large number of modes can propagate, this theory can become very complex and intractable. One will search for simplifications and approximations to the exact theory.

Since many other publications have discussed mode theory of various structures in great detail, (1.11 - 1.14), our approach will be merely to outline the general method and the results. We will then consider two simplifying techniques which give more insight into electromagnetic propagation in dielectric cylinders.

One begins as always with Maxwell's equations. Assuming a linear, isotropic material in the absence of currents and charges, they become

$$\nabla \times \vec{E} = \frac{\partial \vec{B}}{\partial t} \qquad \nabla \times \vec{H} = \frac{\partial \vec{D}}{\partial t}$$

$$\nabla \cdot \vec{B} = 0 \qquad \nabla \cdot \vec{D} = 0 \quad , \tag{1.26}$$

with the constitutive relations,

$$\vec{D} = \varepsilon\vec{E} \qquad\qquad \vec{B} = \mu\vec{H} \quad . \tag{1.27}$$

By taking the curl of the first two equations and applying a vector identity, these can be reduced to the scalar wave equation,

$$\nabla^2 G = \varepsilon\mu \frac{\partial^2 G}{\partial t^2} \quad , \tag{1.28}$$

where, G represents each component of $\vec{E}$ and $\vec{H}$. In doing this, the assumption must be made that $\vec{\nabla}\varepsilon/\varepsilon = \vec{0}$. Marcuse (1.13) shows that if the change in this term is small over a distance of one wavelength this term may be neglected. This is the case for the problems to be considered.

A cylindrical coordinate system r, ψ, and z is defined with the z axis coaxial with the waveguide. Transforming the curl equations to cylindrical coordinates results in two sets of three equations for the components of $\vec{E}$ and $\vec{H}$ in terms of one another. One can solve these for the transverse components E_r, E_ψ, H_r, and H_ψ, in terms of E_z and H_z. We seek solutions which are harmonic in time and z,

$$\begin{bmatrix} E \\ H \end{bmatrix} = \begin{bmatrix} E(r,\psi) \\ H(r,\psi) \end{bmatrix} e^{-i(\omega t - \beta z)} \quad , \tag{1.29}$$

where β is the z component of the propagation vector. With this, the equations for the transverse field components can be written,

$$E_r = \frac{-i}{\kappa^2} \left[\beta \frac{\partial E_z}{\partial r} + \frac{\mu\omega}{r} \frac{\partial H_z}{\partial \psi} \right] \tag{1.30a}$$

$$E_\psi = \frac{-i}{\kappa^2} \left[\frac{\beta}{r} \frac{\partial E_z}{\partial r} - \mu\omega \frac{\partial H_z}{\partial r} \right] \tag{1.30b}$$

$$H_r = \frac{-i}{\kappa^2} \left[\beta \frac{\partial H_z}{\partial r} - \frac{\mu\omega}{r} \frac{\partial E_z}{\partial \psi} \right] \tag{1.30c}$$

$$H_\psi = \frac{-i}{\kappa^2} \left[\frac{\beta}{r} \frac{\partial H_z}{\partial r} + \omega\varepsilon \frac{\partial E_z}{\partial r} \right] \, , \tag{1.30d}$$

where

$$\kappa^2 = k^2 - \beta^2 = \left(\frac{2\pi n}{\lambda} \right)^2 - \beta^2 \, . \tag{1.31}$$

Here k is the propagation constant in a medium of dielectric constant, ε, or alternatively refractive index, n.

The scalar wave equation must now be solved for E_z and H_z to complete the solution. Eq. (1.28) is expressed in cylindrical coordinates and the variables are separated by assuming

$$\begin{bmatrix} E_z \\ H_z \end{bmatrix} = AF(r) \, e^{i\nu\psi} \, . \tag{1.32}$$

From the differential equations for ψ it is found that ν must be an integer to ensure azimuthal periodicity. Later it will be found that $-\mu \leq \nu \leq \mu$, where μ is an integer specifying the number of radial modes.

The differential equation for F(r) becomes

$$\frac{\partial F^2}{\partial r^2} + \frac{1}{r} \frac{\partial F}{\partial r} + \left(k^2 - \beta^2 - \frac{\nu^2}{r} \right) F = 0 \, . \tag{1.33}$$

This equation must be solved for β and $F(r)$ subject to the
boundary conditions of a specific waveguide structure.

1.2.2.2   Homogeneous Core (Step Index) Fiber

1.2.2.2.1 Exact Solution

One of the few refractive index distributions for
which Eq. (1.33) can be solved is that of a homogeneous core of
index n and radius a, surrounded by an infinite cladding of index,
n(1-Δ). This is the step index waveguide considered in the earlier
ray analysis, in which case the solutions are Bessel functions
appropriately chosen to assure finite $F(r)$ at $r = 0$ and $F(r) \rightarrow 0$
as $r \rightarrow \infty$.

For $r < a$, this is a J-type Bessel function of order
ν so that,

$$\begin{bmatrix} E_z \\ H_z \end{bmatrix} = \begin{bmatrix} A \\ B \end{bmatrix} J_\nu(ur) e^{i\nu\psi} \quad , \qquad (1.34)$$

where $u^2 = (k_1{}^2 - \beta^2)$, $k_1 = 2\pi n/\lambda$, and A and B are arbitrary
constants. For the region $r > a$ one must use a modified
Hankel function,

$$\begin{bmatrix} E_z \\ H_z \end{bmatrix} = \begin{bmatrix} C \\ D \end{bmatrix} K_\nu(wr) e^{i\nu\psi} \quad , \qquad (1.35)$$

where $w^2 = \beta^2 - k_2{}^2$, $k_2 = 2\pi n(1-\Delta)/\lambda$, and C and D are again
constants.

The quantity,

$$v^2 = (u^2 + w^2)a^2 = \left(\frac{2\pi a}{\lambda}\right)^2 (2\Delta)n^2, \qquad (1.36)$$

sometimes called the characteristic waveguide parameter, is a
constant of the waveguide and gives much information concerning its operation.

Several observations can be made from Eqs. (1.34) and (1.35). As $wr \to \infty$, $K_\nu(wr) \to e^{-wr}$. For the proper field behavior as $r \to \infty$, $w > 0$. This implies that $\beta \geq k_2$. The equality represents the cutoff condition at which point the propagation is no longer oscillatory and bound to the core region. Inside the core, u must be real, therefore $k_1 \geq \beta$. So we find the allowed range for the propagation constant for bound solutions is

$$k_2 \leq \beta \leq k_1 \quad . \tag{1.37}$$

The exact solution for β must come from satisfying the boundary condition that the tangential components of $\overrightarrow{E}$ and $\overrightarrow{H}$ be continuous at the boundary $r = a$. This condition gives four homogeneous equations in the unknown constants, A, B, C, and D. Only if the determinant of the coefficients vanishes will a solution exist. After much algebra this results in the eigenvalue equation for β,

$$\left[\frac{J_\nu^{'}(ua)}{uJ_\nu(ua)} + \frac{K_\nu^{'}(wa)}{wK_\nu(wa)} \right] \left[\frac{k_1^2 J_\nu^{'}(ua)}{uJ_\nu(ua)} + \frac{k_2^2 K_\nu^{'}(wa)}{wK_\nu(wa)} \right]$$
$$= \nu^2 \beta^2 \left(\frac{1}{u^2} + \frac{1}{w^2} \right)^2 \quad . \tag{1.38}$$

The primes indicate differentiation with respect to the argument. When this equation is solved for β, only discrete values within the range allowed in Eq. (1.37) will be found.

Consider first the case $\nu = 0$. In this case, the fields of the dielectric cylinder break into TM($H_z = 0$) and TE($E_z = 0$) modes just as in the case of the conducting cylinder. Since $\nu = 0$, the modes are radially symmetric. Because of the oscillatory behavior of $J_\nu(ur)$ there will be μ roots of this equation for a given ν, subject to the constraint, $\beta_{\nu\mu} \geq k_2$. These modes correspond to a uniform density of meridional rays, making the same discrete angle with respect to the z-axis.

For $\nu \neq 0$ the situation is more complex. Then hybrid modes, designated $HE_{\nu\mu}$ and $EH_{\nu\mu}$, exist for which both E_z and H_z are non-zero. As before, there exist μ roots for a given ν value. The designation HE or EH is given, depending on whether H_z and E_z make the larger contribution to the transverse field.

An important mode parameter is the cutoff frequency. The following equations give the cutoff conditions for the various mode types:

$$\left.\begin{array}{c} EH_{\nu\mu} \\ \\ HE_{1\mu} \end{array}\right\} J_\nu(u_\mu a) = 0 \qquad\qquad (1.39a)$$

$$HE_{\nu\mu}\left\{(n^2 + 1)J_{\nu-1}(u_\mu a) = \left(\frac{u_\mu a}{\nu-1}\right)J_\nu(u_\mu a) \qquad (1.39b)\right.$$
$$\nu = 2,3\ldots$$

$$\left.\begin{array}{c} TE_{o\mu} \\ \\ TM_{o\mu} \end{array}\right\} J_o(u_\mu a) = 0 \quad . \qquad\qquad (1.39c)$$

There is one mode, designated HE_{11}, for which no cutoff exists. This is the basis for the single-mode waveguide. By adjusting the guide parameters so that the next higher modes, TE_{01}, TM_{01}, HE_{21}, are cutoff, only the HE_{11} is left to propagate. This occurs for

$$2.405 > \frac{2\pi a n}{\lambda} \sqrt{2\Delta} = V \quad . \qquad\qquad (1.40)$$

A plot of the normalized propagation constant β/k, for a few of the low-order modes is shown in Figure 1.4.

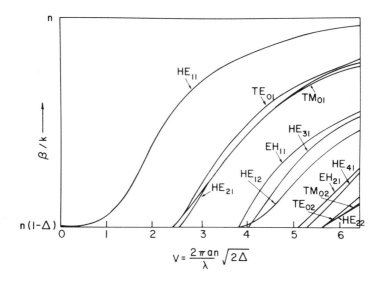

Figure 1.4.   Normalized propagation constant as a function of
               V parameter for a few of the lowest order modes
               of a step waveguide.

Let us next look at the field distribution of modes.
For one polarization, Eq. (1.34) shows the electric field component in the z direction to be

$$E_z \sim J_\nu(u_\mu r)\ \cos\nu\psi \quad , \tag{1.41}$$

whereas the transverse components are obtained from Eq. (1.30),

$$E_r \sim \pm J_{\nu\pm1}(u_\mu r)\ \cos\nu\psi \tag{1.42}$$

$$E_\psi \sim J_{\nu\pm1}(u_\mu r)\ \sin\nu\psi \tag{1.43}$$

where the + and - correspond to the $EH_{\nu\mu}$ and $HE_{\nu\mu}$ modes, respectively. The transverse field is then given by

$$\vec{E}_t = E_r \hat{r} + E_\psi \hat{\psi}$$
$$\sim J_{\nu\pm 1}(u_\mu r)[\pm\cos\nu\psi \ \hat{r} + \sin\nu\psi \ \hat{\psi}] \ . \qquad (1.44)$$

From this, the field patterns of various modes, as well as admixtures of them, can be generated. Since the Cartesian unit vector in the x direction is,

$$-\hat{i} = - \cos \psi \ \hat{r} + \sin \psi \ \hat{\psi} \ , \qquad (1.45)$$

it is seen, for example, that the $HE_{1\mu}$ modes are linearly polarized with the HE_{11} mode being simply proportional to $J_0(ur)$. Also, the $TE_{0\mu}$, $TM_{0\mu}$ modes are seen to be independent of angle, and therefore radially symmetric. In general, the total field pattern obtained will be a complex admixture of the fields of the various modes. This, of course, gives rise to the complex interference pattern observed when coherent light is propagated through a multimode waveguide.

There are no azimuthal zeros in the fields except through linear combinations of modes. Schematically, the first two sets of modes appear as in Figure 1.5.

Although one commonly speaks of a waveguide propagating only the HE_{11} mode as being a single-mode waveguide, this is not strictly correct. Recalling Eq. (1.41), the factor, $\cos\nu\psi$, could also have been taken as $\sin\nu\psi$. This would have generated a transverse field perpendicular to that in Eq. (1.44). Thus the HE_{11} mode is actually a degenerate combination of two mutually orthogonal polarizations.

This degeneracy in the mode propagation constants can be lifted by an azimuthally dependent perturbation (1.15), (1.16). When this is done, extremely coherent propagation in a single

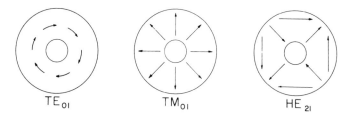

Figure 1.5.   Schematic diagram of the electric field vector for
               the four lowest order modes of a step waveguide.

polarization can take place over very long distances. This fact
allows for many possible uses in integrated optics as well as
various sensors utilizing coherent interference effects, e.g.,
Sagnac (1.17) interferometer gyroscopes, to be considered.

1.2.2.2.2 <u>Weakly Guiding Solution</u>

         The equations just examined are exact solutions for
the homogeneous core waveguide. Snyder (1.18) and Gloge (1.19)
have recognized the fact that because $n \sim n(1-\Delta)$ there is a
great similarity between the eigenvalues for the $HE_{\nu+1,\mu}$ and
the $EH_{\nu-1,\mu}$ modes, which, in fact, are equal if $\Delta = 0$. This
suggests a possible linear combination of solutions to achieve
a simplification.

The assumption is made that for the region $r < a$,

$$E_z = \frac{iAu}{2} \left[J_{\nu+1}(ur) \begin{bmatrix} \sin(\nu+1)\psi \\ \cos(\nu+1)\psi \end{bmatrix} + J_{\nu-1}(ur) \begin{bmatrix} \sin(\nu-1)\psi \\ -\cos(\nu-1)\psi \end{bmatrix} \right]$$

$$H_z = \frac{iAu}{2k} \left(\frac{\varepsilon}{\mu}\right)^{1/2} \left[J_{\nu+1}(ur) \begin{bmatrix} \cos(\nu+1)\psi \\ \sin(\nu+1)\psi \end{bmatrix} - J_{\nu-1}(ur) \begin{bmatrix} \cos(\nu-1)\psi \\ \sin(\nu-1)\psi \end{bmatrix} \right]$$

$$(1.46)$$

As before, $u^2 = k_1^2 - \beta^2$ and $k_1 = 2\pi n/\lambda$. The transverse fields E_r, E_ψ, H_r, and H_ψ are the quantities desired. These may be obtained from Eq. (1.30). It is more instructive to cast them into Cartesian coordinates which may be accomplished by the transformation matrix,

$$\begin{bmatrix} E_x \\ E_y \end{bmatrix} = \begin{bmatrix} \cos\psi & -\sin\psi \\ \sin\psi & \cos\psi \end{bmatrix} \begin{bmatrix} E_r \\ E_\psi \end{bmatrix} . \qquad (1.47)$$

A similar matrix equation exists for the magnetic field. Upon laboriously evaluating the x and y components of $\vec{E}$ and $\vec{H}$ and making the assumption that $n \cong n(1-\Delta)$, it will be found for the assumed E_z and H_z, that $E_x = H_y = 0$ and that

$$E_y = AJ_\nu(ur) \begin{bmatrix} \cos\nu\psi \\ \sin\nu\psi \end{bmatrix}$$

$$H_x = -nA \left(\frac{\varepsilon}{\mu}\right)^{1/2} J_\nu(ur) \begin{bmatrix} \cos\nu\psi \\ \sin\nu\psi \end{bmatrix} . \qquad (1.48)$$

Thus it is seen that this linear combination represents a wave linearly polarized in the y direction. This form of the transverse field is obviously simpler than the exact solutions in Eq. (1.44).

For a complete description of the field, one, of course, requires the mode with orthogonal polarization. This is obtained by making the replacement $\sin(\nu\pm1)\psi \rightarrow \cos(\nu\pm1)\psi$,

$-\cos(\nu\pm1)\psi \rightarrow \sin(\nu\pm1)\psi$ and $J_{\nu-1} \rightarrow -J_{\nu-1}$ for E_z and $\cos(\nu\pm1)\psi \rightarrow$ $\sin(\nu\pm1)\psi$, $\sin(\nu\pm1)\psi \rightarrow -\cos(\nu\pm1)\psi$ and $-J_{\nu-1} \rightarrow J_{\nu-1}$ for H_z in Eq. (1.46). Then it will be found that; $E_y = H_x = 0$, and

$$E_x = AJ_\nu(ur) \begin{bmatrix} \cos\nu\psi \\ \sin\nu\psi \end{bmatrix}$$

$$H_y = nA\left(\frac{\varepsilon}{\mu}\right)^{1/2} J_\nu(ur) \begin{bmatrix} \cos\nu\psi \\ \sin\nu\psi \end{bmatrix} \qquad (1.49)$$

Similar linear combinations are assumed for E_z and H_z in the region $r > a$, except the Bessel functions are replaced by a modified Hankel function of imaginary argument, $K_\nu(wr)$. Upon doing this again, it is found that only linearly polarized fields exist. These have the form,

$$E_y = \frac{AJ_\nu(ua)}{K_\nu(wa)} K_\nu(wr) \begin{bmatrix} \cos\nu\psi \\ \sin\nu\psi \end{bmatrix}$$

$$H_x = -nA\left(\frac{\varepsilon}{\mu}\right)^{1/2} \frac{J_\nu(ua)}{K_\nu(wa)} K_\nu(wr) \begin{bmatrix} \cos\nu\psi \\ \sin\nu\psi \end{bmatrix} \qquad (1.50)$$

$$E_x = H_y = 0$$

for the E_z and H_z in Eq. (1.46), and

$$E_x = \frac{AJ_\nu(ua)}{K_\nu(wa)} K_\nu(wr) \begin{bmatrix} \cos\nu\psi \\ \sin\nu\psi \end{bmatrix}$$

$$H_y = \frac{nAJ_\nu(ua)}{K_\nu(wa)} K_\nu(wr) \begin{bmatrix} \cos\nu\psi \\ \sin\nu\psi \end{bmatrix} \qquad (1.51)$$

$$E_y = H_x = 0$$

for the orthogonal polarization. While these expressions give the form of the field in the various regions of space, we must yet satisfy the boundary condition that the tangential components E_ψ, E_z, H_ψ, and H_z be continuous across the boundary $r = a$. The fields given in expressions (1.48), (1.49), (1.50), and

(1.51) already are such that the components are continuous at
least to the degree that $n \cong n(1-\Delta)$. The continuity of the z
component requires equating like components of $\sin(\nu\pm1)\psi$ and
$\cos(\nu\pm1)\psi$. From this we get the eigenvalue equation which
must be satisfied for a solution to exist:

$$\frac{uJ_{\nu\pm1}(ua)}{J_\nu(ua)} = \frac{wK_{\nu\pm1}(wa)}{K_\nu(wa)} \quad . \tag{1.52}$$

This equation represents a considerable simplification from that given in Eq. (1.38). Snyder (1.18) has shown it
to be accurate within 1 and 10% for $\Delta < 0.1$ and $\Delta < 0.25$,
respectively. We must remember, however, that each solution,
$\beta_{\nu\mu}$, of this equation is really twofold degenerate compared with
the exact solution because of the assumption made concerning the
fields and that $n \cong n(1-\Delta)$. We must also keep in mind that these
field configurations do not really exist at all points along an
actual waveguide since there is always a slight dispersion
between the $HE_{\nu\pm1,\mu}$ and $EH_{\nu\pm1,\mu}$ modes from which they are derived.
This causes these modes to combine alternately in and out of
phase with one another thus never continually exhibiting the
above-field distributions.

The notation for labeling these linearly polarized
modes obviously is no longer the same as for the exact solutions since the integer ν now refers to the combination of exact
modes with labels $\nu+1$ and $\nu-1$. The lowest order mode (HE_{11}) now
has the propagation constant labeled β_{01}. Table 1.1 lists the
relationship between the exact and the linearly polarized modes.

The coefficient A in the expression for the fields
is obtained by evaluating the Poynting vector for a mode. This
is simply the total power in the mode passing through infinite
cross section and thus services to normalize the mode field.

Table 1.1.   Combination of Exact Modes Which Form Linearly
Polarized Modes.

Linearly Polarized	Exact
LP_{01}	HE_{11}
LP_{11}	HE_{21} TE_{01} TM_{01}
LP_{21}	HE_{31} EH_{11}
LP_{02}	HE_{12}
LP_{31}	HE_{41} EH_{21}
LP_{12}	HE_{22} TE_{02} TM_{02}
$LP_{1\mu}$	$HE_{2\mu}$ $TE_{0\mu}$ $TM_{0\mu}$
$LP_{\nu\mu}(\nu \neq 0,1)$	$HE_{\nu+1,\mu}$ $EH_{\nu-1,\mu}$

The assumptions made above become less valid near
cutoff, and since one desires simplified solutions for the
multimode case, it is reasonable to look at approximations far
from cutoffs. It has been shown Marcuse (1.14) in that case,

$$u_{\nu\mu} = u_{\mu}^{\infty} (1- \frac{2\nu}{V})^{1/2\nu} , \qquad (1.53)$$

where u_{μ}^{∞} is the μth root of the equation

$$J_{\nu}(u_{\mu}^{\infty}a) = 0 \quad , \qquad (1.54)$$

and $V = 2\pi a n \sqrt{2\Delta}/\lambda$ is the characteristic guide parameter. For the case of the $HE_{1\mu}$ mode ($\nu = 0$), the limit of Eq. (1.53) simply becomes

$$u_{0\mu} = u_{\mu}^{\infty} \exp(-1/V) \quad . \qquad (1.55)$$

The value of the modal propagation constant, $\beta_{\nu\mu}$, can readily be obtained $\beta_{\nu\mu}^{2} = k_{1}^{2} - u_{\nu\mu}^{2}$.

It is instructive to estimate the total number of modes for a given value of V. This is obtained by counting the total number of roots of Eq. (1.54) for a given value of ν subject to the condition that

$$u_{\mu}^{\infty}a \leq V \quad . \qquad (1.56)$$

It is known that for large μ the roots are, approximately, given by

$$u_{\mu}^{\infty}a \cong (\nu + 2\mu)\pi/2 \leq V \quad . \qquad (1.57)$$

For $\mu = 0$, $\nu = 2V/\pi$; for $\nu = 0$, $\mu = V/\pi$. We see that these two points define a triangle in the ν-μ plane. Each point within the triangle represents four degenerate modes. The total number of modes is therefore given by four times the area of the triangle times the density of modes, $N = 4V^2/\pi^2 = 0.4V^2$. A better approximation has been shown to be (1.19)

$$N = \frac{V^2}{2} \tag{1.58}$$

Finally, we look at the normalized total power for a given mode in the core and cladding regions. This is obtained by integrating the Poynting vector over each region. Gloge (1.19) shows that these are given simply in this field description by

$$\frac{P_{core}}{P_T} = 1 - \frac{u^2}{V^2} \left[1 + \frac{J_\nu^2(ua)}{J_{\nu+1}(ua)J_{\nu-1}(ua)} \right] \, ,$$

$$\frac{P_{clad}}{P_T} = 1 - \frac{P_{core}}{P_T} \, . \tag{1.59}$$

Far from cutoff, Marcuse (1.14) has obtained the expression for the power in cladding,

$$\frac{P_{clad}}{P_T} = \left[\frac{u_\mu^\infty a}{V} \right]^4 \left(1 - \frac{2}{V} \right) \, . \tag{1.60}$$

This clearly shows that as V increases, the fraction of power carried in the cladding for any mode decreases. So, for example, for the HE_{11} mode, one can calculate that for $V = 1$, approximately 70% of the power resides in the cladding while at $V = 2.405$, where the next mode group begins, the situation is reversed and about 84% of the power travels within the core.

The field distribution in the cladding behaves as $K_\nu(wr)$ as seen in Eqs. (1.50) and (1.51). For large r, the asymptotic form is $K_\nu(wr) \rightarrow \exp(-wr)$. Then $r = 1/w$, the field will have decayed to $1/e$ of its maximum value. Defining this as the mode radius, $r_{\nu\mu}$, and applying Eq. (1.53) gives

$$r_{\nu\mu} = \frac{1}{w} = \frac{a}{\left[V^2 - (u_\mu^\infty a)^2 \left(1 - \frac{2\nu}{V} \right)^{1/2} \right]^{1/2}} \, . \tag{1.61}$$

For the HE_{11} mode for example, with $V = 1$, the mode radius is $r_{01} \cong 3a$. One must, therefore, have a cladding on the guide in excess of this to avoid perturbation of the field.

1.2.2.3   Inhomogeneous Core (Graded Index) Fiber

1.2.2.3.1 Exact Solution

Thus far only the homogeneous core waveguide surrounded by an infinite cladding has been considered. There is one other index distribution of significance to optical fibers for which an exact solution to the scalar wave equation can be obtained. That is the square law medium whose refractive index has the form

$$n^2(r) = n^2(o) \left[1 - 2\Delta\left(\frac{r}{a}\right)^2 \right] , \qquad (1.62)$$

where $n(o)$ is the axial index of refraction, Δ is approximately the fractional index difference between core and cladding and, a is the core radius. The solution to this problem has been given Marcuse (1.13) and will not be presented in detail here. The field solutions for this problem are the well-known Laguerre-Gauss functions. The propagation constant for the modes in this case is

$$\beta_{pq} = n(o)k \left[1 - \frac{2\sqrt{2\Delta}}{n(o)ka} \ (p + 2q + 1) \right]^{1/2} \qquad (1.63)$$

It is seen that if $m = p + 2q$, the modes are m-fold degenerate.

1.2.2.3.2 WKBJ Solution

To find solutions to the dielectric waveguide problem that are both simple and applicable to more general refractive index distributions, one can use the WKBJ method, which is well known from quantum mechanics Merzbacher (1.20).

The method has been applied to the solution of the dielectric waveguide problem by Kurtz and Streifer (1.21) and more recently Gloge and Marcatili (1.22). It is a particularly useful technique, especially for obtaining the propagation constant and bears a close examination. It generally ignores anomalies of modes near cutoff.

Recall that if the index distribution is a function only of r, the scalar wave equation (Eq. 1.28) is separable in cylindrical coordinates and has the formal solution given in Eq. (1.32). The differential equation for the radial component of the wave equation must be solved:

$$\frac{d^2G}{dr^2} + \frac{1}{r}\frac{dG}{dr} + \left[k^2(r) - \beta^2 - \frac{\nu^2}{r^2}\right] \tag{1.64}$$

where G represents either E_z or H_z. The radial wavenumber is

$$k(r) = \frac{2\pi n(r)}{\lambda} = k_o n(r) \quad , \tag{1.65}$$

where $n(r)$ is the radial index distribution. As before, ν is the azimuthal mode number and β is the axial component of the propagation vector. The general approach with the WKBJ method is to recognize that if n = constant, the general solution would be a superposition of plane waves

$$G(r) \sim e^{iU(r)} \quad . \tag{1.66}$$

This solution, is substituted into Eq. (1.64), yielding

$$i\frac{d^2U}{dr^2} - \left(\frac{dU}{dr}\right)^2 + \frac{i}{r}\frac{dU}{dr} + k^2(r) - \beta^2 - \frac{\nu^2}{r^2} = 0 \quad . \tag{1.67}$$

If the index variation with r is slow so that the function U(r) is nearly constant over a distance of one wavelength, it can be expanded in a power series in $k^{-1} = (2\pi/\lambda)^{-1}$,

$$U(r) = U_o(r) + \frac{1}{k} U_1(r) + \ldots \qquad (1.68)$$

Substituting this into Eq. (1.67) and gathering like powers of k, to first order, gives the following equations:

$$-\left[\frac{dU_o(r)}{dr}\right]^2 + k^2(r) - \beta^2 - \frac{\nu^2}{r^2} = 0 \quad , \qquad (1.69)$$

$$ik\left[\frac{d^2U_o}{dr^2}\right] - 2\left[\frac{dU_o}{dr}\right]\left[\frac{dU_1}{dr}\right] + \frac{k}{r}\left[\frac{dU_o}{dr}\right] = 0 \quad . \qquad (1.70)$$

Since we shall be concerned primarily with the propagation constant, only the zero order approximation is needed. For a field description $U_1(r)$ would also be required. Integrating Eq. (1.69) gives

$$U_o(r) = \int\left[k^2(r) - \beta^2 - \frac{\nu^2}{r^2}\right]^{1/2} dr \quad . \qquad (1.71)$$

Using the quantum mechanical analog, the quantity $k^2(r) - \nu^2/r^2$ represents a potential well within which a particle of energy β is constrained to move. The potential term, $-\nu^2/r^2$, may be thought of as a centrifugal force and represents the energy associated with the angular motion of the particle. Only if U_o is real will one have an oscillating solution for $G(r)$ and therefore a bound mode. This requires that the radical in Eq. (1.71) be positive. In general, there exist two values r_1 and r_2 for which the radical vanishes. Outside these two values or turning points, U_o becomes imaginary leading to decaying fields.

It can be shown for bound solutions that the phase U_o (evaluated between the turning points) must be approximately a multiple of π,

$$\mu\pi \approx \int_{r_1}^{r_2} \left[k^2(r) - \beta^2 - \frac{\nu^2}{r^2} \right]^{1/2} dr ,$$

$$(1.72)$$

where $\mu = 0, 1 \ldots$ is the radial mode number that counts the number of half periods between the turning points.

Shown schematically in Figure 1.6 is a plot of the radical in Eq. (1.71) or the potential function for the propagating plane waves. Bound modes will be found for any $\beta \geq k_0 n(a)$.

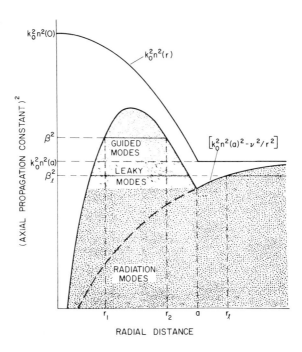

Figure 1.6.   Schematic plot of the waveguide potential as a function of radial distance. The lines $k(r)$ and $k(a)$ represent the core and cladding wave numbers, respectively. β and β_ℓ represent propagation constants for a bound and a leaky mode, respectively.

For a given value of β the meaning of the two turning points is
obvious. It is between these two radii that the ray associated
with the assumed plane wave solution is constrained to move.
For a given β, as ν increases, a point will be reached so that
the turning points merge and beyond which the wave is no longer
bound.

Two other regions are also identified by the potential
well function in Figure 1.6. Although as discussed, bound or
guided modes only exist for $\beta > k_o n(a)$, the propagating field
is still confined to the core region by the centrifugal potential
function, ν^2/r^2 for

$$k_o^2 n^2(a) > \beta^2 > k_o^2 n^2(a) - \frac{\nu^2}{a^2} \quad . \tag{1.73}$$

Modes in this region are referred to as leaky or refract-
ing modes (1.23) since some of the propagating energy can "tunnel"
through the centrifugal potential barrier to the radius r_ℓ, at
which point the wave can radiate from the guide. Generally, the
attenuation coefficient for these modes is somewhat greater than
for bound modes, although as will be shown later many can propagate
over a long distance. Finally, a class of modes exist, which are
not bound to the waveguide, but are part of the radiation mode
continuum and obey the condition

$$\beta^2 > k_o^2 n^2(a) - \frac{\nu^2}{r^2} \quad . \tag{1.74}$$

Using this general picture, a more quantitative
description of the waveguide may be obtained. The number of
bound modes which can exist above a given value of β may be
counted by summing Eq. (1.72) over all ν values,

$$\mu(\beta) = \frac{4}{\pi} \int_{r_1}^{r_2} \int_0^{\nu_{max}} \left[k^2(r) - \beta^2 - \left(\frac{\nu}{r}\right)^2 \right]^{1/2} dr d\nu . \tag{1.75}$$

Here it has been assumed that ν is large enough so that the summation can be replaced by integration. The factor of 4 comes from the degeneracy of the modes with respect to polarization and orientation. To count all the modes, the lower turning point must go to $r = 0$. This will occur for $\nu = 0$. Upon integrating with respect to ν one obtains

$$\mu(\beta) = \int_0^{r_2} [k^2(r) - \beta^2] \, r dr \quad . \tag{1.76}$$

1.2.2.3.3 Power Law Profile

To proceed further with this analysis, specific information regarding the index profile is needed. A particularly useful form is (1.22),

$$n^2(r) = n^2(o)\left[1 - 2\Delta\left(\frac{r}{a}\right)^\alpha\right] , \tag{1.77}$$

where $\Delta = [n^2(o)-n^2(a)]/2n^2(o)$, a is the core radius, and α specifies the shape of the profile. For $\alpha = \infty$ we have simply the step profile, while for $\alpha = 2$ the square law profile results.

The turning point r_2 occurs when $k(r_2) = \beta$. Thus from Eq. (1.77) we obtain

$$r_2 = a\left\{\frac{1}{2\Delta}\left[1 - \left(\frac{\beta}{k_o n(o)}\right)^2\right]\right\}^{1/\alpha} \quad . \tag{1.78}$$

Upon integration of Eq. (1.76) with this upper limit, the number of modes with propagation constants greater than β becomes,

$$\mu(\beta) = \left[\frac{k_o^2 n^2(0) - \beta^2}{2\Delta k_o^2 n^2(0)}\right]^{2+\alpha/\alpha} \left(\frac{\alpha}{\alpha + 2}\right) a^2 k_o^2 n^2(0)\Delta \quad . \tag{1.79}$$

The quantity $\mu(\beta)$ is almost the mode number since it simply counts the modes up to β. We must recall, however, that to this approximation the modes will fail into degenerate groups. Recalling the case $\alpha = 2$, if m is the integer specifying the mode group, each group is m-fold degenerate. Summing the degeneracy of the mode group from zero to the mth level gives the total number of modes above that level, $\mu(\beta)$. Thus one has

$$\sum_{o}^{m} m \approx m^2 = \mu(\beta) \quad . \tag{1.80}$$

All bound modes lie above $\beta = k_o n(a)$, and thus the total number of modes from Eq. (1.79) is

$$N = \left(\frac{\alpha}{\alpha + 2}\right) a^2 k_o^2 n^2(o) \Delta = \frac{V^2}{2} \quad . \tag{1.81}$$

The total number of mode groups is simply $M = \sqrt{N}$. Using this expression with Eq. (1.79) yields the desired expression for the axial propagation constant as a function of the mode group number,

$$\beta_m = k_o n(o) \left[1 - 2\Delta \frac{m}{M}^{2\alpha/\alpha+2} \right]^{1/2} \quad . \tag{1.82}$$

This expression is extremely useful and will be used extensively to discuss dispersion and mode coupling in the waveguide.

The concept of modes and mode group is perhaps made more explicit with the help of Figure 1.7. Schematically shown is the potential well model for three values of the azimuthal mode number, ν. Within the potential well formed for a given value of ν, the energy levels are numbered with the radial mode number, μ, the highest energy having the smallest value. The energy level degeneracies which give rise to mode groups, i.e., levels with different combinations of ν and μ having nearly the same value of β, is illustrated. The mode group numbers follows the relationship

$$m = 2\mu + |\nu| + 1 . \tag{1.83}$$

The total number of modes is simply the phase space volume, and, therefore, measures the light-gathering capability of the waveguide. It is seen from Eq. (1.81) that the total number of modes in the parabolic waveguide, $\alpha = 2$, is one-half of the number of the step waveguide, $\alpha = \infty$. Thus the parabolic waveguide accepts only one-half the light of the step waveguide.

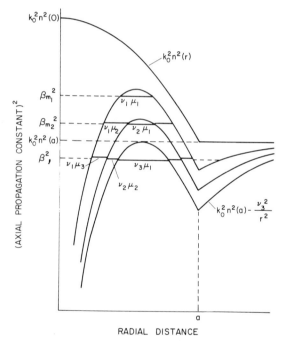

RADIAL DISTANCE

Figure 1.7.   Potential well schematic for a graded-index waveguide showing the relationship between the nearly degenerate levels of axial propagation constant for several azimuthal mode numbers. These degenerate levels form mode groups which are identified by the integer $m = 2\mu + |\nu| + 1$.

Another interesting quantity which will be needed for mode-coupling work involves the mode group spacing, $\delta\beta$. This is given by taking the derivative

$$\frac{d\beta}{dm} = \delta\beta = \left(\frac{\alpha}{\alpha + 2}\right)^{1/2} \frac{2\sqrt{\Delta}}{a} \left(\frac{m}{M}\right)^{\alpha-2/\alpha+2} \qquad (1.84)$$

It is seen that the mode spacing for the step waveguide increases linearly with increasing mode number. On the other hand, for the parabolic waveguide, $\alpha = 2$, the spacing is independent of the mode number as was found earlier.

1.2.3   Ray-Mode Correspondence

1.2.3.1   Waveguide Phase Space

It is desirable to identify more clearly the relationship between the ray and modal descriptions of the waveguide. We again consider the ray described by Figure 1.2 with the spatial variables r, ψ and the angular variables θ, ϕ. With complete generality, the coordinates can be rotated so that $\psi = 0$. The mode propagation constant β is recognized as being the projection of the ray vector k on the axis of the waveguide. The quantity β is a constant of the motion and is conserved for any mode. Mathematically then,

$$\beta_m = k_o n(r) \cos \theta_m(r) \quad , \qquad (1.85)$$

so that for a given group m, the instantaneous angle of the ray θ_m at radial position r depends on the index at that point. By combining Eqs. (1.77) and (1.82), one can easily show that for bound modes, the relative mode group number, m, the radial position and axial angles of the ray, r, and, θ are related:

$$\left(\frac{m}{M}\right)^{2\alpha/\alpha+2} = \frac{\sin^2\theta_m}{2n^2(o)\Delta} + \left(\frac{r}{a}\right)^\alpha \quad . \qquad (1.86)$$

It is noted the mode group number is independent of the azimuthal angle ϕ. This relationship provides a powerful way of explaining waveguide behavior. It relates the general phase space variables $\sin^2\theta$ and r^2 to the mode group number. Shown in Figure 1.8, is such a "phase space" plot for the parabolic waveguide, $\alpha = 2$. It is readily seen that lines of constant relative mode group number m are equally spaced and have a negative unity slope.

For each relative mode group a degenerate set of modes with various relative azimuthal mode numbers ν/M exists. The meridional rays, $\nu/M = 0$, lie in the η-R plane, while helical rays, $\nu/M = 1$, exist at $\eta^2 = \sin^2 \theta/2n(o)^2\Delta = R^2 = (r/a)^2 = 0.5$. This diagram indicates that a ray incident, for example, at $r/a = 0.5$ and $\eta = 0$, will exist at all values $0 < R^2 < 0.5$ and $0 \leq \eta^2 < 0.5$, since the modes in phase space exist not at discrete points but along lines. This explains why a laser beam incident at a specific point on the end face of a graded index waveguide is found within a short length to exist at a range of radial and angular positions which it did not originally possess. This behavior affects measurements of waveguide properties.

The total phase space volume is

$$v = \int_0^a \int_0^{2\pi} \int_0^{\theta_c(r)} \int_0^{2\pi} rdrd\psi \, \sin\theta d\theta d\phi \, , \tag{1.87}$$

which can be integrated, in general, for the α-class profile,

$$v = \left(\frac{\alpha}{\alpha + 2}\right) \pi^2 a^2 n^2(o)\Delta \quad . \tag{1.88}$$

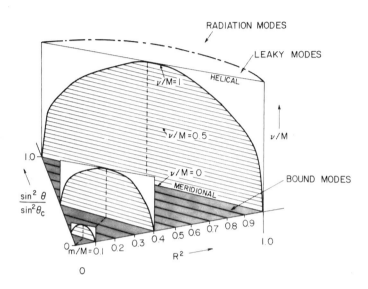

Figure 1.8.   Phase space diagram for a parabolically graded
               index waveguide. The normalized axial angle and
               radius of an incident ray define the degenerate
               mode group in which the light will propagate. The
               azimuthal angle relative to the meridional plane
               dictates which ν value is excited. It is experi-
               mentally found that the attenuation rate and modal
               time delay is nearly constant within a given mode
               group.

     Using Eq. (1.81) gives the expected result for the
volume/plane polarized mode for the waveguide

$$\frac{v}{N/2} = \frac{\lambda^2}{2} \quad .$$ (1.89)

     Thus the modes are uniformly distributed in phase
space with each polarized mode occupying a volume $\lambda^2/2$. The

power input to the waveguide may be calculated by multiplying
the source radiance by the volume/mode, $\lambda^2/2$, and the number
of propagating modes.

It is readily apparent from Figure 1.8 and Eq. (1.81)
that in general, no simple relationship exists between a mode
group number and its ray entry point and angle. However, for
the case of the step profile, $\alpha = \infty$, such a relationship does
exist:

$$\frac{m}{M} = \frac{\sin \theta_m}{n(o) \sqrt{2\Delta}} \quad . \tag{1.90}$$

That is, the far-field pattern of the step waveguide corresponds
exactly to the mode group spectrum of the waveguide. The
general phase-space model for the waveguide presented above,
coupled with differing attenuation and transit times for the
various mode groups, as discussed in the next two sections,
can explain much of the experimentally observed propagation
phenomena of multimode waveguides.

To this point, our ray-mode correlation has focused
on the bound modes. From Eq. (1.85) it can be shown that the
maximum angle that can be accepted by a waveguide and become
a bound mode obeys the relationship

$$0 \leq \sin\theta_{ext} \leq n(o) \sqrt{2\Delta} \left[1 - \left(\frac{r}{a}\right)^\alpha \right]^{1/2} \quad , \tag{1.91}$$

where θ_{ext} is the ray angle external to the waveguide. Of
course the azimuthal mode number is

$$\nu = r \ k \ \sin \theta_{\nu\mu} \sin \phi_\nu \tag{1.92}$$

with the allowed values of ϕ,

$$0 \leq \phi \leq 2\pi \quad . \tag{1.93}$$

The corresponding relationship for leaky modes is,

$$n(0)\ \sqrt{2\Delta}\ \left[1 - \left(\frac{r}{a}\right)^{\alpha}\right]^{1/2} \leq \sin\theta_{ext} \leq n(0)\ \sqrt{2\Delta}\left[\frac{1 - \left(\frac{r}{a}\right)^{\alpha}}{1 - \left(\frac{r}{a}\right)^{2}\sin^{2}\phi}\right]^{1/2}. \quad (1.94)$$

This is schematically illustrated in Figure 1.9. At a given radial position, the bound modes are accepted within the cone of angles given by Eq. (1.91) with no azimuthal dependence. However, leaky modes are only accepted in the shaded regions.

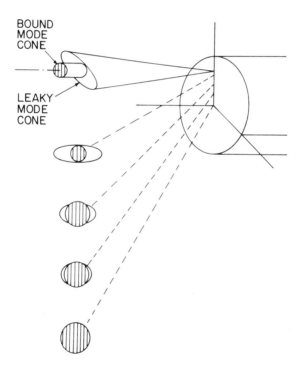

BOUND MODE CONE

LEAKY MODE CONE

Figure 1.9.   Diagram depicting the angular acceptance region of a waveguide depending upon the radius at which the light is incident. Bound modes are accepted within a cone coaxial with the line parallel to the waveguide axis and passing through the entry point. Leaky modes are accepted in the regions outward perpendicularly from the meridional plane as indicated.

1.2.3.2   Near-field and Far-field Distribution

A considerable amount of information about the waveguide can be deduced from the near- and far-field intensity distributions. These are obtained by assuming equal mode excitation and by partially integrating Eq. (1.87) over the angular and radial variable, respectively, subject to the limits in Eqs. (1.91), (1.93), and (1.94). For bound modes, the near-field distribution is

$$\frac{1}{2\pi R} \frac{dP}{dR} = \pi \, a^2 n^2(o) 2\Delta \, (1 - R^\alpha) \quad , \qquad (1.95)$$

where $R = r/a$.

For equal excitation of the modes and ignoring the effect of leaky modes, the near-field distribution is seen to give information about the waveguide profile. Generally, the leaky mode power cannot be ignored, and a correction factor must be incorporated (1.24).

Correspondingly, the far-field distribution is

$$\frac{1}{2\pi \eta} \frac{dP}{d\eta} = \pi \, a^2 n^2(o) 2\Delta \, (1 - \eta^2)^{2/\alpha} \quad , \qquad (1.96)$$

where $\eta = \sin\theta/n(o) \sqrt{2\Delta}$. Again ignoring leaky modes, it is seen that the far-field distribution gives information concerning the maximum angle a wave vector can have with respect to the waveguide axis. For the parabolic profile, this assumption is very good since it can be shown (Problem 10) the leaky modes vanish at the same angle as the bound modes. For other guide types, this assumption is not justified.

1.3   ATTENUATION

1.3.1   General Description

The level of understanding and reduction of attenuation in fiber waveguides is such that it is now possible to
consider transmission links on the order of 10 to 100 km in
length over a fairly broad near-infrared spectral region. The
sources of loss are broadly grouped into absorption and radiation
categories, and can originate from either the material or waveguide
structure.

In general, these two types of loss will be different for the core and cladding regions. We have already seen
in Eq. (1.59) that the power carried in the core and cladding
for the step guide is a function of mode number. Thus if one
assumes that the loss coefficients are γ_1 and γ_2 in the core
and cladding, respectively, the total loss for the $\nu\mu^{th}$ mode
will be

$$\gamma_{\nu\mu} = \frac{\gamma_1 P_{\nu\mu}^{core} + \gamma_2 P_{\nu\mu}^{clad}}{P_{\nu\mu}^{total}} \qquad (1.97)$$

The total loss of the waveguide will obviously be obtained
by summing over all modes weighted by the fractional power in
that mode. The situation for a graded profile guide is far
more complex. Then both the loss coefficient and the modal
power can be functions of the radial coordinate so that one
must integrate over all r to obtain the modal loss prior to
summing over the modes

$$\gamma_{\nu\mu} = \frac{\int_0^\infty \gamma_{\nu\mu}(r) \, P_{\nu\mu}(r) \, rdr}{P_{\nu\mu}^{total}} \qquad (1.98)$$

Techniques now exist for directly measuring the differential attenuation of mode groups (1.25 - 1.27), even though no
exact correlation with an analytical model exists. Utilizing the
phase-space model discussed in the preceding section, Figure 1.10
shows the attenuation as a function of mode group for a typical
multimode waveguide.

The loss, in general, is found to increase with increasing mode group number (1.25). It is readily apparent from this
formulation that the measured attenuation of a waveguide is
strongly dependent upon the manner in which light is injected.

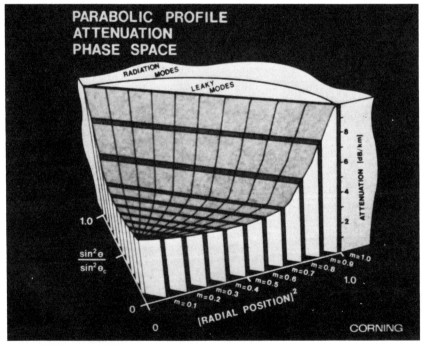

Figure 1.10.   Phase-space attenuation diagram of a parabolic
               profile waveguide. The attenuation rate is found
               to be nearly constant within a given mode group as
               indicated by the vertical planes. High-order mode
               attenuation is often much greater than low-order
               mode attenuation. The attenuation of low-order
               modes is expected to be greater than for interme-
               diate modes because of Rayleigh scattering arising
               from higher dopant concentration on axis.

The rather complicated mechanisms for differential attenuation
are not the primary consideration in this section; rather, the
general loss sources will be discussed with some indication of
their magnitude. The effects of mode coupling will be neglected,
and a discussion of them is postponed to a later section.

1.3.2   Absorptive Loss

1.3.2.1   Intrinsic Absorption
          Absorption loss in glasses can come from three
factors; intrinsic absorption of the basic material, impurity
absorption, and atomic defect absorption. Intrinsic absorption
originates as a result of charge transfer bands in the ultraviolet
region and vibration or multiphonon bands in the near infrared.
If these bands are sufficiently strong, their tails will extend
into the spectral region of interest for fiber communications,
700 to 1700 nm.
          The ultraviolet bands are generally considered the
stronger of the two, but for most glasses presently used in
telecommunications fibers are not a problem. For the case of
germanium-doped silica, Urbach's rule has been applied to the
band edge (1.28) and it was shown that for wavelengths >600 nm,
<1 dB/km absorption resulted.
          Since the initial low-loss waveguides were fabri-
cated, the realization that losses are generally lower toward
longer wavelength has prompted operation in regions where the
infrared vibration absorptions are not negligible. This is a
complete reversal from the earliest days of the technology.
          The wavelength for which the theoretically lowest
intrinsic loss for a pure silica glass occurs is ~1550 nm.
This is at the point where the infrared vibration edge and
the intrinsic scattering (see next section) contribute equally
to the attenuation. For a pure germania glass this wavelength
is about 1750 nm.

It is presently found that waveguides containing boron generally have high attenuation from these infrared bands at wavelengths longer than ~1200 nm. In waveguides containing phosphorus an intrinsic vibration band or a combination vibration with the hydroxyl radical is observed at λ = 1500 nm. Germanium vibration bands are observed to occur beyond about 1700 nm making this a good dopant for waveguides operating at comparatively long wavelengths.

1.3.2.2   Impurity Absorption

Metal ions in glass are traditional sources of impurity absorption. Initially these were most feared, and many studies on bulk glasses showed that the allowed levels for such things as Fe, Cu, V, and Cr could not exceed 8, 9, 18, 8 ppb, respectively, to obtain sub-20 dB/km loss at band center. High silica waveguides are regularly made, however, such as those whose spectra are shown in Figure 1.11 in which these impurities do not contribute to the loss. Although waveguides made from more conventional glasses, for example, using the double crucible process, typically exhibit absorption caused by these impurities, by proper control over the raw material purity and the oxidation conditions during fabrication, losses in the 4 dB/km range can be obtained routinely (1.29). The impurity levels required for low-loss waveguides are such that no direct correlation with waveguide absorption is possible. The only impurity for which a direct correlation has been shown is the OH radical, whose bands at 1380, 1250, and 950 nm are clearly visible in Figure 1.11. These are overtones and combination bands of the fundamental OH vibration at 2.73 μm and of silica matrix. The strength of the 950 nm band has been shown to be ~1 dB/km/ppm. Waveguides can be made routinely with <0.1 ppm.

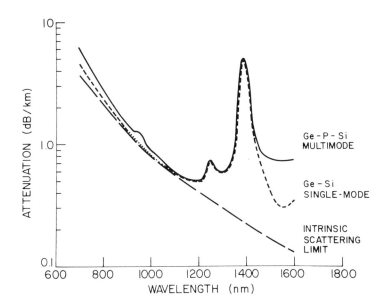

Figure 1.11.   Attenuation spectrum for a high silica waveguide
between 700 and 1600 nm. The dotted curve indicates the total scattering. Clearly visible are
the OH absorption bands.

1.3.2.3   Induced Absorption

Atomic defect absorption is induced in the glass by
an external stimulus. Examples of such stimuli are intense
radiation or large temperature variations. The magnitude of
such induced losses can be quite large as, for example, in the
case of titanium-doped silica where a reduction $Ti^{4+} \rightarrow Ti^{3+}$,
occurs during the high-temperature fiberization to produce
losses of several thousand dB/km (1.30). A similar band occurs
in silica glass at ~630 nm (1.31) but is much less severe.

Similarly, defects induced by high intensity radiation, e.g. gamma, neutron, or electron radiation, can cause extremely high losses in optical waveguide glasses (1.32). Many factors affect the magnitude of the loss at a given wavelength (1.33) - glass composition, type and strength of the radiation, the duration of the radiation, sample temperature, and hydroxyl content of the glass. The origin of this loss is presently thought to arise from nonbridging oxygen bonds, either in the glass matrix itself or as a result of impurities such as alkali, transition metal, aluminum, or hydroxyl ions. Exposure of such a matrix to ionizing radiation creates electrons and holes which can be trapped at the defect sites. These trapped electrons and holes generally absorb radiation in the blue region of the spectrum. However, the absorption bands can be sufficiently intense and broad that they affect the light propagation in the near-infrared spectral region ($\lambda \geq 800$ nm), which is of concern for optical communication. Of course, the radiation, if sufficiently energetic, can create the defect sites by physically displacing ions within the glass. In addition to causing absorption centers, light is generated either as Cerenkov radiation or as luminescence when the electrons and holes recombine.

A bleaching of this radiation-induced loss occurs even at room temperature, thus the attenuation increase as a function of dose is somewhat reduced by this effect. Under pulsed radiation, the glass attenuation increases to a peak value and then bleaches to its original state in a few seconds. This again depends on the glass type. For example, under ^{60}Co irradiation a germanium-borosilicate waveguide may absorb 1 to 3 dB/km and 40 to 50 dB/km at doses of 100 and 2000 rads, respectively (1.34), depending on the exact composition and possibly the OH content of the glass. This loss will decay to the original state in about one second. A much slower recovery is observed for germanium-phosphosilicate waveguides.

1.3.3   Radiative (Scattering) Loss

1.3.3.1   Intrinsic (Rayleigh) Scattering

All transparent materials scatter light as a result of thermal fluctuations of the constituent atoms that are frozen in during solidification. These cause density and hence index variations within the material leading to intrinsic scattering. This is believed to represent the fundamental limit to attenuation in waveguides. The intrinsic scattering can be calculated by subdividing the sample into small volumes which act as dipoles. One then sums the contributions of all dipoles integrated over all angles and can relate the scattering loss to the isothermal compressibility, B,

$$\gamma_s = \frac{8\pi^3}{3\lambda^4} (n^2-1) \; kTB \quad , \tag{1.99}$$

where T is the transition temperature at which the fluctuations are frozen into the glass. This loss decreases rapidly with increasing wavelength. For fused silica using a transition temperature of $1500°C$, one calculates a loss of 1.7 dB/km at 820 nm which is in good agreement with experiment. A tradeoff between the transition temperature and the compressibility for a given material has been shown to exist (1.35). In fact, lithium-aluminosilicate glasses have been measured to have losses caused by density fluctuations several times less than that of fused silica.

For waveguides, one generally has glasses with more than one oxide, in which case another form of scattering occurs. This, scattering is a result of concentration fluctuations in the constituent oxides and also causes a loss. The expression for this loss is

$$\gamma_s = \frac{16\pi^3 n}{3\lambda^4} \left(\frac{dn}{dc}\right)^2 (\Delta c)^2 \delta V \quad , \tag{1.100}$$

in which $(\Delta c)^2$ is the mean square concentration fluctuation
and δV is the volume over which it occurs. Generally, the
magnitude of the index fluctuation is not known, and, therefore,
the scattering cannot be calculated. Rather, the scattering is
used to obtain the index fluctuation. It is generally observed
that if the added oxide raises the refractive index, larger
fluctuations tend to occur. Thus for high index glasses,
losses caused by concentration fluctuations occur. However,
for high silica glasses, concentration fluctuations typically
account for only ~25% of the observed scattering loss. It
should be noted that these concentration fluctuations increase
directly as the dopant concentration increases (1.36). This
then becomes one design tradeoff, for as the dopant concentration is increased to raise the waveguide numerical aperture,
or input coupling efficiency for incoherent sources, the
attenuation rate is simultaneously increasing as the square
of the numerical aperture. This provides a maximum distance
over which a benefit is obtained by increasing input coupling.
For example, increasing the numerical aperture from 0.2 to 0.3
theoretically permits an additional length of ~1.2 km to be used
for a 3 dB/km waveguide.

1.3.3.2   Induced Scattering

          In addition to these two intrinsic scattering loss
mechanisms, one can induce scattering through nonlinear effects,
such as both forward and backward stimulated Raman and Brillouin
scattering. Because of the small core size, the confined guidance and the long interaction length, relatively low absolute
power levels are required to observe such effects (1.37). For
long lengths of fiber, expressions for the maximum power at
1.06 μm for these two processes have been derived (1.38),

$$P \sim 1 \times 10^{12} \ A\alpha \ (\text{watts})$$
$$(\text{Backward Raman})$$

(1.101)

$$P \sim 5 \times 10^9 \ A\alpha \ (\text{watts})$$
$$(\text{Backward Brillouin}),$$

where $A(\text{cm}^2)$ is the power carrying area and $\alpha(\text{cm}^{-1})$ is the loss
coefficient. These expressions predict, for example, that
threshold powers of ~225 and 1.1 W would be required for Raman and
Brillouin scattering, respectively, for a 50 μm core diameter
fiber with a 4 dB/km attenuation. At these thresholds, the fiber
loss would show a rapid increase due to the rapid dissipation of
energy by these two processes. The above equations have assumed
the pump to the scattering process spectral bandwidth.

Generally, this will not be the case, $\Delta\nu$ (Brillouin)
< $\Delta\nu$ (Raman) < $\Delta\nu$ (laser), and hence considerably higher thresh-
olds will occur. In fact, 2000 have been injected into a 2 km
length of guide having a 75 μm core diameter with no nonlinear
attenuation observed (1.39).

In the backward scattering case, we have been concerned with a direct waveguide loss. Forward Raman scattering
has also been observed in both multimode (1.40) and single mode
waveguides (1.41). Here, although energy is not lost, it is up
or down shifted to other wavelengths. Since these typically
travel at different velocities within the waveguide, an interference effect could arise relative to information transmission.
Fortunately, for multimode waveguides the threshold power density
is extremely high, ~200 MW/cm^2, and need not be considered. For
single-mode waveguides much lower powers are required, but it is
still not expected to be a problem. At present, the single-mode
fiber Raman laser is being used as a measurement tool in many
laboratories (1.41).

1.3.4   Waveguide Structure Loss

1.3.4.1   Finite Cladding
            In addition to the above scattering loss mechanisms
radiation losses can be associated with the intrinsic waveguide
structure. We have assumed the cladding to be infinitely thick
while in practice it is on the order of a few tens of microns.
Thus if the jacket is lossy, as, for example, to minimize crosstalk,
some fraction of the mode fields can reach this region and be
attenuated. If n_j is the index of the jacket region which occurs
at a distance $r \geq r_j$, then the rate of radially outward power flow
relative to axial power flow is simply

$$N' = \left[\frac{(kn_j)^2 - \beta^2}{\beta^2} \right]^{1/2} . \qquad (1.102)$$

            Multiplying this by the relative power at the cladding-
jacket interface gives the loss coefficient for the $\nu\mu^{th}$ mode,

$$\gamma_{\nu\mu}(dB) = 4.34\pi \ r_j \ \frac{p_{\nu\mu}(r_j)N'}{p_{\nu\mu}^{total}} , \qquad (1.103)$$

where $p(r_j)$ is the power density at the interface and $p_{\nu\mu}^{total}$
is the total power in the mode. To obtain the total loss, all
modes must be summed over. This is a very complex computation
(1.42), (1.43). Gloge (1.42) has obtained an approximate
expression for the cladding thickness necessary to restrict the
loss of a fraction, F, of modes to greater than 1 dB/km. He
finds

$$r_j \cong a\left[1 + \frac{36}{\sqrt{FV^3}}\right] . \qquad (1.104)$$

            For a typical multimode guide with V = 50, a jacket
radius of $r_j \cong 1.3a$ would be required for 90% of the modes to

have loss less than 1 dB/km. Typical waveguides today are designed with a core radius of 25 μm and a cladding radius of 62 μm, more than fulfilling this requirement.

1.3.4.2   Leaky Mode Loss

Thus far only bound modes have been considered in our discussions. The restriction for these modes, $\beta \geq n_2 k$, strictly holds only for $\nu = 0$. However, as discussed previously using the potential well model for $\nu \neq 0$, a mode with $\beta < n_2 k$ only turns purely radiative beyond the radius

$$r_\ell = \nu[(kn_2)^2 - \beta_\ell^2]^{-1/2} . \tag{1.105}$$

The large ν, the leakage of the evanescent field through the region, $a < r < r_\ell$, can be small, and hence these modes can propagate long distances. Pask, et.al., (1.44) have derived an expression for the loss coefficient for these modes for large ν, in the step index fiber,

$$\gamma_\ell = \frac{4}{\pi a \sqrt{2\Delta}} \frac{\theta_z^2}{V} \frac{1}{|K_\nu(wa)|^2} , \tag{1.106}$$

where $\theta_z = \cos^{-1}(\beta/n_1 k)$. Upon evaluating this expression, a number of modes will be found with a loss coefficient $\gamma_\ell < \gamma_0$. An approximate expression for the fractional increase in mode volume is (1.42),

$$f \approx 0.1 \left[\frac{a\gamma_0}{\sqrt{\Delta}}\right]^{1/2} . \tag{1.107}$$

For a typical multimode guide with $\Delta = 0.01$ and $a = 25$ μm, for $\gamma_0 = 1$ dB/km this would represent a fractional increase in mode volume of ~7%. The total number of leaky modes is given by (1.44).

$$N_\ell = \frac{V^2}{2} \left[1 - \frac{8}{3\pi} \sqrt{2\Delta} + \ldots \right] . \qquad (1.108)$$

For small Δ, the number of leaky modes represents a significant fraction of the number of guided modes and consequently can have a significant effect on propagation.

1.4   INFORMATION CAPACITY

1.4.1   General Description

In this section the effects of waveguide dispersion will be considered. This will be done primarily from the standpoint of digital transmission in which case pulse broadening produced by the waveguide is of primary concern. Later sections of this book will formally relate the amount of pulse broadening to the information carrying capacity of a waveguide transmission system.

The approach will be to first obtain a general expression for pulse broadening, and then to apply the results of the WKBJ section to evaluate the result for the case of the power-law profile. It will be found that for a certain radial index profile, pulse broadening can be made very small giving rise to large information capacity. However, fundamental restriction placed by material dispersion will always limit the information capacity. The complication of mode coupling between modes will be ignored here and postponed until the next section.

1.4.2   Intermodal and Intramodal Pulse Broadening

1.4.2.1   Theory

It has been seen that the modes of a waveguide are specified by two integers ν and μ which enumerate the azimuthal and radial modes of the electromagnetic fields. The propagation constant $\beta_{\nu\mu}$ is found to depend on the core radius, a,

and the propagating wavelength, λ. For considering pulse
propagation the group velocity or, alternatively, the group
delay per unit length for a given mode is required:

$$\tau_{\nu\mu} = \frac{1}{v_g} = \frac{1}{c} \frac{d\beta_{\nu\mu}}{dk} , \qquad (1.109)$$

where $k = 2\pi/\lambda$ is the free space propagation constant. If the
waveguide is excited by a pulse $g(t)$, its response at the point
z, for a given wavelength λ, will be obtained by a summation
over all modes,

$$P(t,z,\lambda) = \sum_{\nu\mu} P_{\nu\mu}(z = o,\lambda)\, e^{-\gamma_{\nu\mu}(\lambda)z}\, g[t - \tau_{\nu\mu}(\lambda)z] . \qquad (1.110)$$

The functions $P_{\nu\mu}(0,\lambda)$, $\gamma_{\nu\mu}(\lambda)$ and $\tau_{\nu\mu}(\lambda)$ are the
power at wavelength λ input to mode $\nu\mu$ at $z = 0$, the attenua-
tion at λ of mode $\nu\mu$, and the group delay per unit length at λ
as given by Eq. (1.109), respectively. The quantity $P_{\nu\mu}$ in
theory, is obtained from an overlap integral between the incident
field radiance distribution B, and the field distribution of
the $\nu\mu^{th}$ waveguide $E_{\nu\mu}$,

$$P_{\nu\mu} = \frac{\displaystyle\int B(r,\psi;\theta,\phi)\, E_{\nu\mu}(r,\psi;\theta,\phi)dv}{\displaystyle\int \left| E_{\nu\mu}(r,\psi;\theta,\phi) \right|^2 dv} . \qquad (1.111)$$

For the discussion of pulse broadening, the source
radiance B will be considered a constant resulting in equal
modal excitation.

While this is not strictly true for many sources
used with waveguides, the result obtained by this analysis is
mathematically tractable and will in fact represent a worst
case result. Similarly, as shown in Section 1.3, the attenuation $\gamma_{\nu\mu}$ varies with mode number. However, after sufficient

distance, only modes with similar low attenuation will remain. Therefore, the assumption of constant modal attenuation will also be made.

These assumptions, in addition to ignoring the effects of mode coupling, considerably simplify Eq. (1.110) but will nevertheless give a reasonably good picture of pulse propagation in the waveguide. Furthermore, since most detectors are incapable of resolving individual wavelengths the quantity of practical interest is

$$P(t,z) = \int P(t,z,\lambda)d\lambda \quad . \tag{1.112}$$

A brief digression is required at this point. A parameter is needed to specify the amount of pulse broadening associated with the pulse distribution $P(t,z)$. For systems analysis the Fourier transform provides perhaps the most useful tool,

$$\tilde{P}(\omega,z) = \int_{-\infty}^{\infty} P(t,z)\, e^{-i\omega t}dt \quad . \tag{1.113}$$

Alternatively, however, the moments $M_n(z)$ of the distribution $P(t,z)$ defined by the equation

$$M_n(z) = \int_{-\infty}^{\infty} t^n P(t,z)dt \tag{1.114}$$

can do the same task. Often knowledge of only the first few moments provides sufficient information and hence this specification is considerably simpler. From Eqs. (1.113) and (1.114) one can show that

$$M_n(z) = i^n \frac{d^n}{d\omega^n}\tilde{P}(\omega,z)\bigg|_{\omega=0} \quad . \tag{1.115}$$

Hence the complete set of moments is completely identical to specifying the Fourier transform.

If $g(t)$ is taken as an impulse $\delta(t)$, applying the assumptions above with the moment definition, and integrating over all wavelengths, Eq. 1.110 becomes

$$M_n(z) = z^n \int_{-\infty}^{\infty} \sum P_{\nu\mu}(\lambda,z) \, \tau^n_{\nu\mu}(\lambda) d\lambda \quad . \qquad (1.116)$$

The main wavelength dependence in the mode power will come from the source spectral power distribution, $S(\lambda)$, which is generally sharply peaked. It, therefore, can reasonably be assumed that

$$P_{\nu\mu}(\lambda,z) \cong S(\lambda) P_{\nu\mu}(z) \quad . \qquad (1.117)$$

Using the general definition of moments, the following equations may be written:

$$S_0 = \int_0^{\infty} S(\lambda) d\lambda \qquad (1.118a)$$

$$\lambda_0 = \frac{1}{S_0} \int_0^{\infty} \lambda S(\lambda) d\lambda \qquad (1.118b)$$

$$\sigma_s^2 = \frac{1}{S_0} \int_0^{\infty} (\lambda - \lambda_0)^2 \, S(\lambda) d\lambda \quad , \qquad (1.118c)$$

where $S_0 = M_0$ is the total source power, $\lambda_0 = M_1$ is the mean operating wavelength, and $\sigma_s = M_2$ is the rms spectral source width.

If the group index is a smoothly varying function in the vicinity of λ_0, then the group delay $\tau_{\nu\mu}$ may be expanded in a Taylor series,

$$\tau_{\nu\mu}(\lambda) = \tau_{\nu\mu}(\lambda_0) + \tau'_{\nu\mu}[\lambda - \lambda_0] + \tau''_{\nu\mu}[\lambda - \lambda_0]^2 + \cdots \quad . \qquad (1.119)$$

The primes denote the derivative with respect to λ evaluated at λ_0. Substituting this into Eq. (1.116) and utilizing the definitions in Eq. (1.118), gives an expression for the nth moment

$$M_n(z) \simeq z^n \sum P_{\nu\mu}(z) \left\{ \tau_{\nu\mu}^n(\lambda_0) + \frac{\sigma_s}{2\lambda_0^2} \left[n\tau_{\nu\mu}^{n-1}(\lambda_0) \; \lambda_0^2 \; \tau_{\nu\mu}'' \right. \right.$$
$$\left. \left. + n(n-1)\tau_{\nu\mu}^{n-2}(\lambda_0) \; \lambda_0^2 (\tau_{\nu\mu}')^2 \right] \right\} \; , \tag{1.120}$$

since $\sigma_s/\lambda_0 \ll 1$, terms on the order of $(\sigma_s/\lambda_0)^3$ and higher may be neglected. This expression may be more simply written if we define $\langle A \rangle$ as the average of the variable A over the mode distribution:

$$\langle A \rangle = \sum P_{\nu\mu}(z) A_{\nu\mu}/M_0 \; . \tag{1.121}$$

Then a normalized moment m_n can be defined,

$$m_n = \frac{M_n}{M_0} = z^n \left\{ \langle \tau^n(\lambda_0) \rangle + \frac{\sigma_s^2}{2\lambda_0^2} \left[n\lambda_0^2 \langle \tau^{n-1}\tau'' \rangle \right. \right.$$
$$\left. \left. + n(n-1) \; \lambda_0^2 \langle \tau^{n-2} (\tau')^2 \rangle \right] \right\} \; . \tag{1.122}$$

With this definition, $m_0 = 1$. The average arrival time $\tau(z)$ is simply given,

$$\tau(z) = m_1 = z \left[\langle \tau(\lambda_0) \rangle + \frac{\sigma_s^2}{2\lambda_0^2} \langle \lambda_0^2 \tau'' \rangle \right] \; . \tag{1.123}$$

The quantity in square brackets is the reciprocal group velocity for the guide structure.

For the purpose of specifying the information carrying capacity, the second moment or rms pulse width $\sigma(z)$ is most useful (1.45). With the above definitions it becomes

$$\sigma^2(z) = m_2 - m_1^2 = \sigma^2_{\text{intermodal}} + \sigma^2_{\text{intramodal}} \; , \tag{1.124}$$

where,

$$\sigma^2_{intermodal} = z^2 \left\{ \langle \tau^2(\lambda_o) \rangle - \langle \tau(\lambda_o) \rangle^2 \right. \tag{1.125}$$
$$\left. + \left(\frac{\sigma_s}{\lambda_o} \right)^2 \left[\langle \lambda_o^2 \tau''(\lambda_o) \tau(\lambda_o) \rangle - \langle \lambda_o^2 \tau''(\lambda_o) \rangle \langle \tau(\lambda_o) \rangle \right] \right\}$$

and

$$\sigma^2_{intramodal} = z^2 \left(\frac{\sigma_s}{\lambda_o} \right)^2 \langle \lambda_o^2 \tau'(\lambda_o)^2 \rangle \; . \tag{1.126}$$

The rms width has been separated into two components
for the purpose of clarifying the physical situation. Both
are dependent on the length of the waveguide. The intermodal
broadening, Eq. (1.125), results from differences in the group
delay between the various modes. The portion of this term
that depends on the relative source spectral width is small
and can usually be neglected. The intramodal term, Eq. (1.126),
represents the average pulse broadening within each mode and
variously goes by the terms "material dispersion" or "chromatic
dispersion." This term provides the only nonvanishing dispersion contribution for a single-mode guide and thus defines the
ultimate limitation on information capacity. It contains two
distinct parts, one arising from the bulk material, the other
from the guide structure. The separation can be made explicit
by writing the mode delay time,

$$\tau_{\nu\mu} = \frac{N}{c} + \delta\tau_{\nu\mu} \quad , \tag{1.127}$$

where N is the group index, N/c represents the delay common to
all modes, and $\delta\tau_{\nu\mu}$ is the correction introduced by the guide
structure. Then, taking the derivative of Eq. (1.127) with
respect to wavelength gives

$$\tau'_{\nu\mu} = -\lambda n'' + \delta\tau'_{\nu\mu} \quad , \tag{1.128}$$

where n is the guide refractive index. Inserting this into
Eq. (1.126) yields

$$\sigma^2_{intramodal} = z^2 \left(\frac{\sigma_s}{\lambda_o}\right)^2 \left[(\lambda_o n'')^2 - 2\lambda_o^2 n'' \langle \lambda_o \delta\tau' \rangle \right.$$
$$\left. + \langle (\lambda_o \delta\tau')^2 \rangle \right]. \tag{1.129}$$

The first term is a pure material effect, the last a
pure waveguide effect, and the middle a mixed material-waveguide
term.

For multimode waveguides the latter two terms are of
minor importance since the intermodal term tends to dominate.
However, as the number of propagation modes is reduced, these
terms become nonnegligible. In particular, for the single-mode
waveguide it has been shown that fairly large changes in
wavelength response of $\sigma_{intramodal}$ are possible (1.46-1.48).
Considering the multimode case for the moment, the intramodal
term is simply,

$$\sigma^2_{intramodal} = z \left(\frac{\sigma_s}{\lambda_o}\right) \lambda_o^2 \frac{d^2 n}{d\lambda^2} . \tag{1.130}$$

This term defines the ultimate information capacity
limit for all types of waveguides. It can be reduced by
decreasing the relative source spectral width. Alternatively
for a given glass composition, it may be possible to operate
at wavelengths for which n'' is reduced.

1.4.3   Power Law Profile Pulse Broadening

To carry the analysis further, expressions for the
group delay must be obtained. This can be done most simply
using the results of the WKBJ solution obtained in the previous
section. For the mth mode group the propagation constant β_m is
given in Eq. (1.82). Considering both the explicit and implicit

k-dependence, the group delay is given from Eq. (1.109),

$$\tau_m = \frac{NL}{c}\left\{1 + \Delta\left[\frac{\alpha - 2 - \varepsilon}{\alpha + 2}\right]\left(\frac{m}{M}\right)^{2\alpha/\alpha+2}\right.$$

$$\left. + \Delta^2\left[\frac{3\alpha - 2 - 2\varepsilon}{2(\alpha + 2)}\right]\left(\frac{m}{M}\right)^{4\alpha/\alpha+2}\right\} + O\Delta^3 \quad , \tag{1.131}$$

where

$$\varepsilon = -\frac{2n}{N}\frac{\lambda}{\Delta}\frac{d\Delta}{d\lambda} \quad . \tag{1.132}$$

The quantity, ε, takes into account the varying dispersion properties between the core and cladding materials (1.49).

From this it is seen that if $\alpha > 2 + \varepsilon$, higher order mode groups will have a greater relative delay. The opposite is true of $\alpha < 2 + \varepsilon$. To first order in Δ, the group delay difference between modes vanishes $\alpha = 2 + \varepsilon$. Since ε is generally small, this indicates that the near parabolic index profile tends to minimize the intermodal dispersion. It is recalled from the earlier ray propagation discussion that the parabolic profile resulted in a near focusing condition. It is thus seen that the focusing and minimum dispersion condition are synonomous.

Recalling the relationship in Eq. (1.86) between the mode group number, the axial angle, and radial position of a ray, a phase space picture of the modal delay time may also be given as shown in Figure 1.12. It is obvious from this pictorial representation that the resulting value of σ for a waveguide can depend dramatically on the source energy distribution launched. Laser sources which tend to produce low radius and moderate angle excitation will have much different pulse propagation characteristics than LED sources which more uniformly irradiate the waveguide phase space.

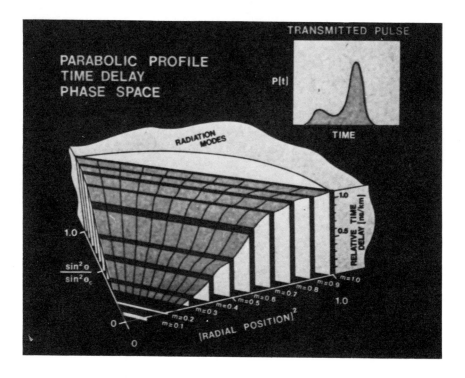

Figure 1.12.   Phase space relative time delay diagram for a
               parabolic profile waveguide. Delay times are
               found to be nearly constant for a given mode
               group as suggested by the theory. Both positive
               and negative relative delay time can be obtained
               depending on the profile shape.

Approximating the summation in Eq. (1.121) with an
integration and assuming equal exitation of the modes, the
intermodal and intramodal contributions to the rms width are

$$\sigma_{intramodal} = \frac{LN\Delta}{2c}\left(\frac{\alpha}{\alpha+1}\right)\left(\frac{\alpha+2}{3\alpha+2}\right)^{1/2}\left[c_1^2 + \frac{4C_1C_2\Delta(\alpha+1)}{2\alpha+1}\right.$$
$$\left. + \frac{4\Delta^2c_2^2(2\alpha+2)^2}{(5\alpha+2)(3\alpha+2)}\right]^{1/2} \tag{1.133}$$

and

$$\sigma_{intramodal} = \frac{L\sigma_s}{c\lambda} \left[\left(\lambda \frac{d^2n}{d\lambda^2} \right)^2 - 2\lambda^2 \left(\frac{d^2n}{d\lambda^2} \right)^2 N\Delta C_1 \left(\frac{\alpha}{\alpha+1} \right) \right.$$

$$\left. + (N\Delta)^2 c_1^2 \left(\frac{2\alpha}{3\alpha+2} \right) \right]^{1/2} \qquad (1.134)$$

where

$$C_1 = \frac{\alpha-2-\varepsilon}{\alpha+2}$$

$$(1.135)$$

$$C_2 = \frac{3\alpha-2-2\varepsilon}{2(\alpha+2)} \qquad .$$

The minimum intermodal delay is found to occur for

$$\alpha_o = 2 + \varepsilon - \Delta \frac{(4+\varepsilon)(3+\varepsilon)}{5+2\varepsilon} \qquad . \qquad (1.136)$$

1.4.3.1 Material and Profile Dispersion

The correction to α_o involving Δ in Eq. (1.136) occurs because of a partial cancellation of the mode dependent term in Eq. (1.133). The quantities $(\lambda/c)(d^2n/d\lambda^2)$ and ε are known to depend on both composition and wavelength, and thus can substantially alter the spectral pulse transmission characteristics of the waveguide.

A summary of this dependence is shown in Figures 1.13 and 1.14 (1.50-1.53).

All materials exceed the material dispersion value for silica except phosphorus which in this wavelength range is nearly the same. It is also noted that at ~1260 nm, $(\lambda/c)(d^2n/d\lambda^2)$ passes through zero. This is the so-called zero dispersion wavelength, at which point the material dispersion vanishes leaving only the waveguide term in Eq. (1.129).

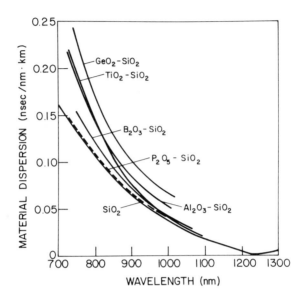

Figure 1.13.   Plots of material dispersion for several different
               silicate glasses. All glasses have a dispersion
               greater than that of pure SiO₂. The material dis-
               persion of all these glasses goes through zero in
               the range 1270 nm to 1320 nm making this a favorable
               region in which to operate.

     The other materials shown also exhibit this zero cross-
ing but at somewhat longer wavelengths. For example, in Ge-Si
waveguides, this occurs at ~1300 nm. For sources with broad
spectral bandwidths, σ_s, there is a distinct advantage in operating at this zero dispersion point as well as shown in latter
graphs.
     Since the quantity, ε, referred to as "profile disper-
sion," is a function of wavelength and composition (Figure 1.14),
the value α_0 for the optimum waveguide profile will also be a

function of wavelength. Alternatively, for a given profile, the rms pulse broadening will be a function of wavelength. This has been demonstrated experimentally many times (1.49). Again it is noted that since phosphorus has a dispersion quite close to that of silica, its optimum profile parameter exhibits a small change with wavelength.

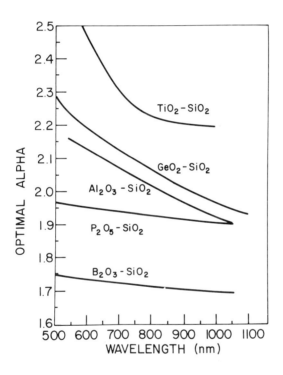

Figure 1.14.   Plots of the optimum profile parameter as a function of wavelength.   At 900 nm the binary Ge-Si
glass optimum profile is parabolic.

1.4.3.2   Information Capacity

      Considering a Ge-Si waveguide operation at 900 nm,
Figure 1.15 shows the information carrying capacity calculated
from Eqs. (1.133) and (1.134) for three different GaAs light
sources. Here the conversion between bit rate and rms pulse
broadening,

$$B \cong \frac{0.2}{\sigma} \quad , \tag{1.137}$$

has been used (1.45). This relationship will be discussed in
Chapter 6.

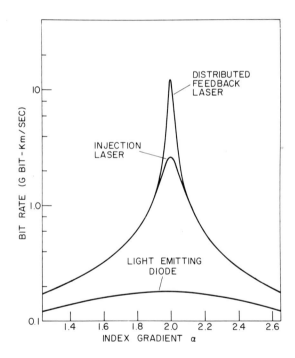

Figure 1.15.   Plot of the waveguide information-carrying capa-
               city as a function of profile parameter α. In
               each of these curves the peak bit rate is limited
               by material dispersion associated with the source
               spectral bandwidth.

The curves are for an LED, an injection laser, and
a distributed feedback laser which might typically have rms
spectral widths of 150 Å, 10 Å, and 2 Å, respectively. The
effect of diminishing the source spectral width is readily
apparent near $\alpha = 2$, since the intramodel dispersion dominates
there.

It is also noted that the fiber profile for this
given wavelength must be fairly well controlled around the
optimum profile in order to achieve high information carrying
capacity. At the 1 GBit-km/sec level, the profile tolerance
is about 5%. This obviously gets much more severe at higher
bandwidth values.

1.4.4   Profile Perturbations

Even if the actual average profile shape is at the
optimum value, perturbations about the correct value will
cause a reduction in the bit rate which can be transmitted.
The resulting rms pulse broadening being obtained by taking
the theoretically perfect broadening and the perturbation
broadening in quadrature,

$$\sigma^2_{actual} = \sigma^2_{theory} + \sigma^2_{pert} \qquad . \qquad (1.138)$$

Using first-order perturbation theory one can show
that the mean square deviation from the optimum shape averaged
over the guide length determines the magnitude of σ_{pert} and
hence the waveguide information carrying capacity reduction.

The effect has been calculated for a gaussian shaped
bump (1.54), a "stairstep" (1.55) approximation, as well as a
sinusoidal oscillation (1.54) about the correct value. The
effect can be severe. For example, with a 1% amplitude sinusoidal
variation at the worst possible frequency, an approximately
100 MBit-km/sec limiting bit rate is imposed on the waveguide.

These perturbations are fortunately not present in this magnitude
for most practical waveguides.

A comment should be made concerning the dependence of
the information capacity on the relative core-cladding index
difference, Δ. As shown in Eq. (1.133) the intermodal broadening
increases directly as Δ for $\alpha \neq \alpha_o$. In the of $\alpha = \alpha_o$ region,
the broadening becomes proportional to Δ^2. Thus a tradeoff will
exist between increasing Δ to obtain better source coupling and
decreasing Δ to reduce intermodal dispersion.

Let us consider the wavelength response of a waveguide
in somewhat more detail. Using the data presented in Figures
1.13 and 1.14 and Eqs. (1.133) and (1.134) the graph in Figure
1.16 may be calculated. Here a Ge-Si composition is assumed.
Since the achievable losses are very low in the 1200 to 1500 nm
region, one is also interested in the dispersion characteristics
in this region. Correspondingly, the assumption was made in
Figure 1.16 that the waveguide profile was tuned for 1100 nm.
The source is assumed to correspond to a laser with $\sigma_s \cong 0.1$ nm.
At 1100 nm, the bit rate is theoretically limited by the material
dispersion of the glass to less than $\sim$30 GBit-km/sec. If we had
chosen a wavelength of $\cong 1300$ nm, near the material dispersion
zero, the source bandwidth would have played a negligible role.
Indeed many system designs are incorporating LEDs operating at
this wavelength to make use of this fact. It was mentioned
earlier that in both step and graded index core single-mode
waveguides when waveguide and material dispersion are added,
the wavelength at which the total dispersion zero occurs can
be dramatically shifted (1.46), (1.47), (1.41), (1.48), (1.56),
and (1.57).

Closed-form approximations for the magnitude of this
correction have proved to be insufficiently accurate to predict
the shifts. Instead, numerical solutions for the particular

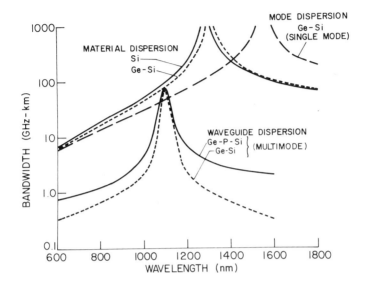

Figure 1.16.   Bandwidth as a function of wavelength for both
               single- and multimode doped Si waveguides. The
               material dispersion curve provides the upper limit
               on bandwidth for both types. By combining mode
               with material dispersion in a properly designed
               single-mode waveguide this limit may be shifted
               toward longer wavelength as indicated. Profile
               dispersion in the multimode waveguide causes the
               bandwidth to reach the material limit at only a
               single wavelength. The width of the multimode
               dispersion curve depends on the waveguide compo-
               sition.

profile and guide parameters being examined must be used. One

can, however, construct single-mode waveguides for which the

total dispersion can be made to vanish at any wavelength between

those shown in Figure 1.16.

1.4.5   Pulse Broadening Wavelength Dependence

Returning to graded index multimode guides, as the source wavelength departs from the value for which the guide profile is tuned, the bit rate decreases dramatically. This has been experimentally observed in several waveguides (1.58). The effect of waveguide perturbations will still be as shown schematically by the dashed curve in Figure 1.16. Although approximately three orders of magnitude improvement can theoretically be obtained in an optimally graded relative to a simple step profile, the highest recorded values are ~3 GHz-km (1.59).

Since the sources, which will be described in Chapter 4, can be made to emit in the 800 to 900 nm as well as the 1100 to 1700 nm wavelength ranges, and since the waveguide attenuation discussed in Section 1.3 is also sufficiently low in these regions, considerable thought has been given to broadening the wavelength region over which the waveguide profile is optimized. Two alternatives have been posed:

o   Utilize glass compositions for which $d\alpha/d\lambda$, Figure 1.14 is small, so that a single profile shape is optimum (1.60).

o   Utilize several different compositions each with appropriately optimized profiles (1.61), (1.62).

Several variations on these ideas have also been proposed (1.63-1.65).

The first of these methods is based on the fact that the optimum profile parameter for multicomponent glasses is given for small Δ by

$$\alpha(\lambda) = 2 + \frac{1}{\Delta} \sum_i \Delta_i [\alpha_i(\lambda) - 2] \quad , \qquad (1.139)$$

where Δ_i is the fractional contribution of each material to the total Δ, and $\alpha_i(\lambda)$ is optimum profile at a given wavelength for

the ith component. From the curves in Figure 1.14, one can show
that an extremely high phosphorus and small germanium content, in
principle, would give rise to a very flat wavelength dependence
of the profile parameter. In practice, this is difficult to fabricate; however, some movement in this direction is possible,
and waveguides with >1 GBit-km/sec at both 900 and 1300 nm have
been observed (1.66).

1.4.6   Pulse Shape

1.4.6.1   Theory

To this point, the width of the transmitted pulse has
been the primary concern. However, for system calculations,the
exact shape of the pulse is also of importance. As indicated
in Eq. (1.110), this, in general, will depend strongly upon
differential mode attenuation as well as the specific excitation
conditions of the waveguide. However, if uniform modal excitation is assumed and no differential attenuation exists, then a
rather simple expression for the pulse shape can be obtained.
This gives a reasonable qualitative description and thus warrants
examination. By inverting Eq. (1.131) and retaining only terms
of order Δ, an expression for the mode group number in terms of
its relative group delay, $\delta\tau$, is obtained:

$$m = M \left[\frac{\alpha + 2}{\alpha - 2 - \varepsilon} \frac{1}{\Delta} \right]^{\alpha+2/2\alpha} (\delta\tau)^{\alpha+2/2\alpha} \qquad . \qquad (1.140)$$

The relative amount of power δP arriving per unit
time is simply

$$\delta P = \left[\frac{2m}{M^2} \frac{dm}{d(\delta\tau)} \right] = \left[\frac{\alpha + 2}{\alpha - 2 - \varepsilon} \frac{1}{\Delta} \right]^{\alpha+2/\alpha} \left(\frac{\alpha + 2}{\alpha} \right) (\delta\tau)^{2/\alpha} \qquad (1.141)$$

The above expression holds only for $\alpha \neq 2 + \varepsilon$. As this condition is violated, higher order terms in Δ must be retained. Eq. (1.141) is plotted in Figure 1.17. For $\alpha = \infty$, the arriving power is independent of the relative delay resulting in a gate function. The width of the gate, obtained as the difference between the earliest and latest mode arrival time from Eq. (1.131), is exactly as would be calculated from a meridional ray analysis, $\Delta t = LN\Delta/c$. For $\alpha \neq \infty$, since the higher order modes carry a larger fraction of the energy, the pulse is peaked at the relative delay of the highest order mode. Thus for $\alpha > 2 + \varepsilon$ this occurs later than the delay for the lowest order mode. The opposite is true for $\alpha < 2 + \varepsilon$.

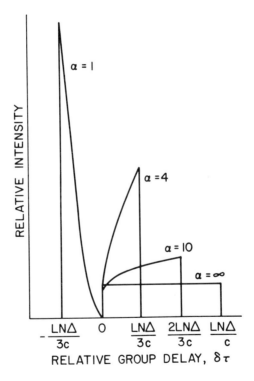

Figure 1.17.   Relative pulse shape as a function of the index gradient parameter α (1.22).

1.4.6.2   Optical Equalization

The above switch in the relative mode arrival time, depending upon which side of the optimum profile the actual fiber profile lies, affords another way for achieving high information carrying capacity in long-distance systems. This is the concept of "optical equalization" (1.67). Generally, for a concatenation of I fibers, the theorem on variance states that the rms pulse broadening is given by

$$\sigma^2 = \sum_{i=1}^{I} \sigma_i^2 + \sum_{i=1}^{I} \sum_{\substack{j=1 \\ i \neq j}}^{I} \sigma_i \sigma_j r_{ij} \quad , \tag{1.142}$$

where r_{ij} is a correlation coefficient between the relative modal delay time of the i and j fibers. Ignoring the effects of material dispersion, to first order in Δ the rms width of a transmitted pulse is given by Eq. (1.133):

$$\sigma = \frac{NL\Delta}{2c} \left(\frac{\alpha}{\alpha + 1} \right) \left(\frac{\alpha + 2}{3\alpha + 2} \right)^{1/2} \left(\frac{\alpha - \alpha_o}{\alpha + 2} \right) \tag{1.143}$$

where α_o is the optimum profile for that guide. Assuming all I guides have the same composition (and therefore α_o), length, and profile (near $\alpha = 2$), this reduces to

$$\sigma_i = \frac{NL\Delta}{12\sqrt{2}c} (\alpha_i - \alpha_o) \quad . \tag{1.44}$$

For uncorrelated behavior, i.e., random and uniform coupling of mode power to every other mode, $r_{ij} = 0$. In this case the total concatenated pulse broadening for I fibers is

$$\sigma = \sqrt{I} \left(\frac{NL\Delta}{12\sqrt{2}c} \right) \left[\overline{(\alpha_i - \alpha_o)^2} \right]^{1/2} \quad , \tag{1.145}$$

where $\overline{(\alpha_i - \alpha_0)^2}$ is the average squared deviation of the fiber profiles from the optimum. It is noted that the total rms pulse broadening increases only as the square root of the total number of fibers or system length. The potential advantage of this case will be discussed more completely in Section 1.5.3.3 when waveguide perturbations are considered.

For the case of correlated behavior, i.e., zero mode coupling, the situation is more complex. However, assuming the fibers are not too close to the optimum profile, the total rms pulse width becomes

$$\sigma = I \left(\frac{NL\Delta}{12\sqrt{2}c} \right) \overline{(\alpha_i - \alpha_0)} \quad . \tag{1.146}$$

Here it is noted the broadening increases directly as the number of fibers or total system length, but more importantly, it shows that by combining fibers that are alternately above and below the optimum (negatively correlated), the total broadening may be kept very low. This effect has been demonstrated (1.68) but relies heavily on coherent propagation with respect to individual mode groups over long lengths. This may be difficult to achieve in practice. All systems are expected to fall between the length-dependent behavior forecast by Eqs. (1.145) and (1.146).

To complete this section on pulse propagation, we return again to the correspondence between the modal and ray picture for a step waveguide. Using Eq. (1.86), the group delay, Eq. (1.131), can be rewritten in terms of the mode angle,

$$\tau_m = \frac{LN}{c} \left[1 + \frac{\alpha - 2 - \varepsilon}{\alpha + 2} \left(\frac{\theta^2}{2} \right) + \frac{3\alpha - 2 - 2\varepsilon}{\alpha + 2} \left(\frac{\theta^4}{8} \right) + \cdots \right] . \tag{1.147}$$

As already stated, the correspondence between mode number and angle is strictly only good for the step guide, in which case this equation becomes

$$\tau_m = \frac{LN}{c} \left[1 + \frac{\theta^2}{2} + \frac{3\theta^4}{8} + \cdots \right] = \frac{LN}{c} \sec\theta \quad . \tag{1.148}$$

It is seen that this is the same as Eq. (1.4) if the latter is multiplied by the group velocity of the guide. Thus for the step waveguide, a meridional ray analysis gives a correct picture for intermodal pulse broadening.

1.5   PROPAGATION IN NON-IDEAL FIBER WAVEGUIDES

1.5.1   General Description

Thus far, the characteristics of the intrinsically perfect waveguide have been considered. However, it is highly unlikely that some form of perturbation will not exist in the waveguide either as a result of the processes by which it is fabricated or more likely as a result of its environment. These perturbations can take the form of "intrinsic" perturbations of the waveguide refractive index and/or core diameter or "extrinsic" perturbations such as deviation of the fiber axis from straightness. In any case, they will be assumed to vary with position along the waveguide. It will be seen that these perturbations can have a significant effect on both the information carrying capacity as well as the waveguide attenuation.

For these longitudinally varying perturbations, a power spectrum of spatial frequencies at which they occur can be defined. The propagating rays in a perturbed waveguide can easily be envisioned to reflect off these periodic perturbations in some new direction. If the new direction is characteristic of a different allowed bound ray for the waveguide, mode coupling occurs. The situation however, can be that the ray merely adiabatically alters its characteristics, remaining in the same basic mode. This can be put on a more mathematical basis using Eq. (1.84). There an expression for the difference in propagation constants between adjacent modes in a power-law profile waveguide was given,

$$\Delta\beta_m = \left(\frac{\alpha}{\alpha+2}\right)\frac{2\sqrt{\Delta}}{a}\left(\frac{m}{M}\right)^{\alpha-2/\alpha+2} .$$

(1.149)

This may be thought of as a separation of the waveguide energy levels. Generally speaking, if the perturbation function, $f(z)$, possesses a frequency component in its power spectrum,

$$F(\omega) = \frac{1}{\sqrt{L}} \int_0^L f(z)e^{i\omega z} \, dz$$

(1.150)

at the value, $\Delta\beta_m$, coupling of energy can occur between mode m and $m \pm 1$. If, however, the components of the power spectrum are at too low a frequency (energy), then mode coupling will not occur. Nevertheless, even in this case, the transmission properties of the guide are altered. Two regions can, therefore, be defined dependent on the frequency components in $F(\omega)$,

$$\omega > \Delta\beta_{min} = \frac{\alpha}{\alpha+2} \frac{2\sqrt{\Delta}}{a} \left(\frac{1}{M}\right)^{\alpha-2/\alpha+2}$$

(1.151)

(mode coupling region)

and,

$$\omega < \Delta\beta_{min} \qquad \text{(adiabatic transfer region)} .$$

(1.152)

For a typical parabolic waveguide with a = 25 μm and
Δ = 0.01 the critical spatial wavelength is

$$\Lambda = \frac{2\pi}{\Delta\beta} \sim 1.1 \text{ mm} \quad , \qquad (\alpha = 2) \qquad (1.153)$$

while for a step waveguide with the same core radius and fractional index difference, coupling will occur over the frequency
range,

$$1 \text{ mm} \leq \Lambda \leq 18 \text{ mm} \qquad (\alpha = \infty) \quad . \qquad (1.154)$$

The two cases will be considered in some detail.

1.5.2   Low-Frequency Perturbations

Within this region two separate cases can also be
identified. Perturbations that are longitudinally stationary,
i.e., have zero spatial frequency, can affect propagation. One
such case, for example, was considered in the previous section
where the effect which a radial index profile perturbation had
on pulse broadening was discussed. This profile perturbation
was assumed to be constant along the length and nevertheless
decreased the waveguide information capacity. This particular
type of perturbation had no effect on attenuation.

1.5.2.1   Waveguide Curvature

Another such zero frequency perturbation which can be
achieved readily is a constant curvature of the waveguide axis.
This is sometimes referred to as a macrobend in contrast to a
microbend perturbation, which will be considered subsequently in
the mode-coupling discussion. This type of perturbation produces
an attenuation of selected modes, and recalling Eq. (1.110), can

also be expected to affect pulse propagation. The physical
picture for macrobend loss is that as the modal field is constrained to move along a curved path, in order to maintain
coherency across its wavefront the mode phase velocity will be
forced to exceed the velocity of light beyond some critical
radius (1.69). This forces radiation of the mode power to occur
in order to not violate the laws of physics. Experimental veri-
fication of this model has been obtained (1.70). With simple
geometric pictures one can show this critical radius to be

$$r_c = R \left[\frac{\beta}{n(a)k_o} - 1 \right] \quad . \qquad (1.155)$$

The macrobend attenuation rate is obtained by comparing
the total modal power beyond the critical radius with the total
propagating modal power. Modal power, of course, redistributes
constantly as the loss occurs. Several theoretical calculations
of this loss have been made (1.71-1.73).

This may alternatively be looked at using the potential
well model of Figure 1.18 where it has been shown (1.74) that the
effect of the bend is to conformally transform the waveguide
potential,

$$V(r) = \left[1 + \frac{r}{R} \cos\psi \right] [n(r) - \left(\frac{\nu}{r} \right)^2] \quad . \qquad (1.156)$$

Here the angle, ψ, is the azimuthal angle measured
from the center of the fiber and the plane of the bend. This
clearly shows macrobending to be a tunneling phenomenon.

The theories described above show that for a given
bend radius, the modal attenuation is very low until a critical mode group number is reached beyond which the loss rate is
extremely high. Therefore, after a sufficient length, all
groups beyond m_c are lost. Combining Eqs. (1.82) and (1.155),
taking the core radius for r_c, and neglecting terms of order
Δ^2, yields the expression for the critical mode group,

$$\frac{m_c}{M} \cong \left[1 - \frac{a}{R\Delta}\right]^{\alpha+2/2\alpha} \tag{1.157}$$

This represents a transient loss of power,

$$\gamma_B \cong 10 \ \log \left[1 - \frac{m_c}{M}\right] \tag{1.158}$$

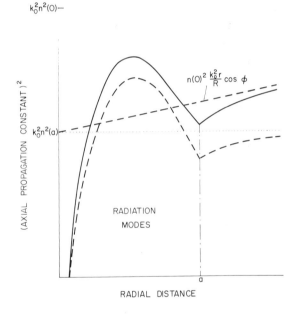

Figure 1.18.   Potential well model of a curved waveguide.
Waveguide curvature corresponds to increasing
the effective index linearly in the radial direction by a term inversely proportional to the bend
radius. Thus the potential well is rotated upward
at increasing radius thereby allowing high order
modes to more easily tunnel out of the waveguide
and radiate.

As long as a/RΔ is small, i.e. R $\gtrsim$ 5 cm for typical telecommunication waveguides (a ~25 μm, Δ = 0.01), the loss can be written

$$\gamma_B \cong 10 \ \log \left[\frac{\alpha + 2}{2\alpha} \quad \frac{a}{R\Delta} \right] . \qquad (1.159)$$

It is noted that the information capacity can be altered by this perturbation since the delay time of only those mode groups less than m_c will contribute to the pulse broadening.

Although this is admittedly a simplistic model of macrobending loss, it does demonstrate the general findings of a more rigorous analysis. The loss of the parabolic guide, α = 2, for example, is seen to be twice as large as that of the step guide, $\alpha = \infty$. Further macrobending losses are decreased by increasing Δ and decreasing the core radius of the guide. While this loss mechanism is not a serious practical problem for most long-distance deployments of optical fiber, it is a considera-tion for splice points and also for some connector designs where fibers may be coiled in small spaces.

1.5.2.2   Diameter Variations

Core diameter and index fluctuations are expected to be nonzero but nevertheless are low frequency perturbations as defined by Eq. (1.152). Such perturbations can produce loss since previously propagating modes may become leaky modes as the potential well gradually shifts in response to the pertur-bation. A random diameter variation as small as 2% has been calculated to produce a 0.56 dB transient loss in a 1 km waveguide (1.75). With present manufacturing tolerances, however, this and similar variations in refractive index do not seem to be a problem.

1.5.3 High Frequency (Mode Coupling) Perturbations

1.5.3.1 Mode Coupling Theory

Perhaps the most physically interesting propagation phenomena occur when the waveguide is subjected to high frequency longitudinal perturbations. In this case coupling of power between modes occurs.

Generally, Maxwell's equations must be solved for the mode fields subject to the specific distortion of the waveguide. This involves solutions of an infinite set of coupled differential equations which are usually very difficult to solve. The solution also contains more information than is often desired. A considerable simplication can be obtained by considering only the coupled power flow (1.76) in the guide. Consider a guide having M modes. The total change in power $P_m(z,t)$ of the mth mode group, as a function of distance will be

$$\frac{dP_m}{dz} = \frac{\partial P_m}{\partial z} + \frac{\partial t}{\partial z}\frac{\partial P_m}{\partial t} = \frac{\partial P_m}{\partial z} + \frac{1}{v_m}\frac{\partial P_m}{\partial t} \quad . \qquad (1.160)$$

where v_m is the group velocity of the mode group. This change in power is caused by three factors:

o   the power dissipated from the mode through normal loss and scattering

o   the power lost to other modes through coupling

o   the power gained from other modes through coupling.

The coupled power flow equation may then be written,

$$\frac{\partial P_m}{\partial z} + \frac{1}{v_m}\frac{\partial P_m}{\partial t} = -\gamma_m P_m + \sum_{n=1}^{M} d_{mn}(P_m - P_n) \quad , \qquad (1.161)$$

where d_{mn} is the coupling coefficient from the mode group m to
n and is assumed reciprocal, and γ_m is the normal loss coefficient of the mode group. This represents M coupled differential
equations which, in general, have no analytic solution. To
obtain a simplication, coupling is assumed to be dominated by
transitions for which n = m ± 1. This adjacent mode group coupling is quite reasonable for most practical cases. Additionally
the modes are assumed to be very closely spaced so that a transformation to a single continuous variable may be made. If the
average coupling coefficient between the m and m ± 1 mode group
is d(m), then Eq. (1.160) may be rewritten,

$$\frac{\partial P(m)}{\partial z} + \frac{1}{v(m)} \frac{\partial P(m)}{\partial t} = - \gamma(m)P(m) + \frac{1}{m} \frac{\partial}{\partial m} \left[md(m) \frac{\partial P(m)}{\partial m} \right].$$

(1.162)

This equation is seen to be a diffusion equation for
the power in the mth mode group. Many solutions to this equation are available in the literature for various assumptions
concerning the parameters v(m), γ(m) and d(m) (1.76), (1.74),
(1.77), (1.78), and (1.79). Among these are included power law
and gaussian models for d(m) and modal losses, which are both
constant or increase as a power of the mode number. These have
been done for single-mode (1.80), (1.81), and (1.79), as well
as multimode waveguides. All are based, in part, on practical
assumptions. However, such wide variations in the actual conditions can exist that no model can account for all cases. In the
following discussion, one of these sets of assumptions will be
made and the solution found in order to elucidate propagation in
this type of perturbed guide.

    '   For the class of index profiles in Eq. (1.77), the
group velocity for the mth mode group can be obtained from Eq.
(1.131). A reasonable assumption for γ(m) based on experimental
evidence is that

$$\gamma(m) = \gamma_0 \qquad\qquad m < M_c$$
$$\gamma(m) = \infty \qquad\qquad M_c < m < M(\alpha) \quad , \qquad (1.163)$$

where M_c allows for the fact that only some fraction of the total number of mode groups, $M(\alpha)$, Eq. (1.81), might propagate. These restrictions give one boundary condition which will be needed for solution of Eq. (1.162),

$$P(m) = 0 \qquad\qquad m > M_c \quad . \qquad (1.164)$$

The other boundary condition is obtained from the fact that the "diffusion current" at $m = 0$ is zero,

$$I(0) \equiv m \, d(m) \left. \frac{\partial P(m)}{\partial m} \right|_{m=0} = 0 \quad . \qquad (1.165)$$

This, of course, represents the fact that no power can flow to modes with $m < 0$.

Derivation of the coupling coefficient is very complex (1.14), and here we merely outline the main parts. Assume the refractive index distortion can be expressed by the fairly general function,

$$\delta n^2(r,\psi,z) = g(r) \cos(\ell\psi+\xi) \, f(z) \quad , \qquad (1.166)$$

where $f(z)$ contains all the longitudinal dependence and $g(r)\cos(\ell\psi+\xi)$ contains the transverse dependence. Then the coupling coefficient for a large number of small distortions can be simply expressed,

$$d(m) = |D(m)|^2 \, \langle |F(\Delta\beta_m)|^2 \rangle \quad . \qquad (1.167)$$

In doing this, we must insist that the departure of the waveguide from the ideal symmetry is either small or slow compared with wavelength of the propagating light. The quantity, $D(m)$, depends only on the transverse index distortion and has the form

$$D(m) \sim D_{mm'} \sim \int \vec{E}_m^* \cdot \vec{E}_{m'} g(r) \cos(\ell\psi + \xi) r \, dr \, d\psi \quad , \qquad (1.168)$$

where $\vec{E}$ is the transverse electric field vector for the mth mode group of the particular waveguide type being considered. It is seen that $D_{mm'}$ is a measure of the degree to which the distortion causes an overlap of the fields of the m and m' mode groups. It is this quantity which also determines the selection rules for mode coupling. This quantity may be exactly calculated using the $\cos(\ell\psi+\xi)$ distortion for the step ($\alpha = \infty$) and the parabolic ($\alpha = 2$) profiles. Using the WKBJ method for the index profile in Eq. (1.77), an expression for $D(m)$ may be developed which linearly interpolates between these two results for an arbitrary α,

$$D(\alpha, m) = \frac{nka}{2\sqrt{2}} \left(\frac{m}{M}\right)^{2/\alpha+2} (\Delta\beta)^2 \quad . \qquad (1.169)$$

The quantity $F(\Delta\beta_m)$ is the Fourier transform of the longitudinal distortion function $f(z)$ over the length of the guide,

$$F(\Delta\beta_m) = \frac{1}{\sqrt{L}} \int_0^L f(z) \, e^{-i\Delta\beta_m z} \, dz \quad . \qquad (1.170)$$

If the function $f(z)$ contains no spatial frequencies corresponding to the separation, $\Delta\beta_m \equiv \beta_m - \beta_{m'}$, between the m and m' mode groups, then no mode coupling will occur. Obviously a sinusoidal longitudinal dependence will couple only a single-mode group spacing. As seen in Eq. (1.84) the mode group spacing depends on the type of waveguide.

All mode groups of the parabolic guide ($\alpha = 2$) can be coupled with a single frequency, whereas a spectrum of frequencies is required for the step guide ($\alpha = \infty$). If Eq. (1.170) is integrated twice by parts, it is mathematically equivalent to

$$F(\Delta\beta) = \frac{1}{\Delta\beta^2} \frac{1}{\sqrt{L}} \int_0^L \frac{d^2 f}{dz^2} e^{-i\Delta\beta z} dz \quad .$$

(1.171)

The quantity $d^2 f/dz^2$ is the curvature of the waveguide axis whose distortion is given by $f(z)$. If $C(\Delta\beta)$ is the Fourier transform of this curvature, then

$$F(\Delta\beta) = \frac{1}{\Delta\beta^2} C(\Delta\beta) \quad .$$

(1.172)

The quantity $\langle F(\Delta\beta)^2 \rangle$ then is seen to be the average power spectrum of either the distortion of the waveguide axis or its curvature. Thus by combining Eqs. (1.167), (1.169), and (1.172) we have the coupling coefficient for the case of random curvatures of the waveguide axis. The curvature function is specified by $C(\Delta\beta)$.

Even though the function $d(m)$ can be specified, Eq. (1.162) can be solved in terms of standard functions only for certain types of distortion. One such form which is fairly general and has physical significance is

$$d(m) = d_o \left(\frac{m}{M}\right)^{-2q} \quad .$$

(1.173)

For $q = 0$ the coupling has the constant value d_o.

1.5.3.1.1 <u>Steady State Solution</u>.

Before considering pulse behavior of mode coupling, insight can be obtained by analyzing the case where $\partial P/\partial t = 0$. With the preceding assumptions on $\gamma(m)$ and $d(m)$, Eq. (1.162) can be written

$$\frac{\partial P}{\partial z} = - \gamma_o P + \frac{d_o}{M_c^2} \frac{1}{\chi} \frac{\partial}{\partial \chi} \chi^{1-2q} \frac{\partial P}{\partial \chi} , \qquad (1.174)$$

where the normalized group number $\chi = m/M_c$ has been introduced. The solution to this equation is (1.77),

$$P_j(\chi,z) = P_j(\chi) e^{-(\gamma_o + \gamma_j)z} , \qquad (1.175)$$

where,

$$P_j(\chi) = N_j \chi^q J_\ell(\lambda_j \chi^{1+q}) \qquad (1.176)$$

and

$$\gamma_j = \frac{d_o}{M_c^2} (1 + q)^2 \lambda_j^2 . \qquad (1.177)$$

The order of the Bessel function is $\ell = \pm q/1+q$. The boundary condition, Eq. (1.164), that no power can exist at $\chi > 1$ gives

$$J_\ell(\lambda_j) = 0 . \qquad (1.178)$$

This condition defines the jth value of λ, which is simply the jth zero of J_ℓ, $Z_j(\ell)$. The normalization factor N_j is obtained from the other boundary condition, Eq. (1.165),

$$N_j = (1 + q)^{\frac{1}{2}} \left| J_{\ell+1} (Z_j(\ell)) \right|^{-1} . \qquad (1.179)$$

The most general solution to Eq. (1.162) is a superposition of the solutions in Eq. (1.175),

$$P(\chi,z) = e^{-\gamma_o z} \sum_{j=1}^{\infty} I_j P_j(\chi) e^{-\gamma_j z} , \qquad (1.180)$$

where the coefficients, I_j, are determined from the initial
power distribution, $I(\chi)$, at $z = 0$,

$$I_j = \int_0^{\infty} 2\chi I(\chi) P_j(\chi) d\chi \qquad (1.181)$$

Eq. (1.180) shows that the power, $P(\chi,z)$, is attenuated
by the normal loss coefficient, γ_o. Additionally, it is made up
of a linear combination of modes whose power distributions, $P_j(\chi)$,
are each attenuated by an excess loss coefficient, γ_j, which
depends on the magnitude of the coupling coefficient, d_o. Since
γ_j is proportional to the square of the jth zero of the ℓth Bessel
function, and since $Z_j(\ell)$ increases with j, it is seen that for
large z the term $\exp(-\gamma_j z)$ rapidly attenuates all terms above
$j = 1$. Thus the steady-state mode coupled power distribution
will quickly approach

$$P(\chi,z) = I_1 P_1(\chi) e^{-(\gamma_o + \gamma_1)z} . \qquad (1.182)$$

Thus for a step fiber with constant coupling between
mode groups, i.e., $q = 0$, the steady-state distribution will
be

$$P(\chi,z) \sim J_o(2.405\chi) e^{-d_o\left(\frac{2.405}{M_c^2}\right)^2} . \qquad (1.183)$$

For a step guide for which $\chi = m/M_c = \theta/\theta_c$, the far-field angular distribution will exhibit this power distribution.

1.5.3.1.2 Time Dependent Solution

When the proper dependence of $v(m)$ is inserted into Eq. (1.162) no analytic solution is possible. Even a computer calculation of the power distribution for a specific case is a complex problem. However, a great deal of information as well as a considerable simplification can be obtained by considering the moments of the impulse response as defined in Eqs. (1.114) and (1.115).

By introducing the Laplace transform,

$$R(\chi,z,s) = \int_0^\infty P(\chi,z,t)\,e^{-st}\,dt \quad , \tag{1.184}$$

Eq. (1.162) may be transformed to,

$$\frac{\partial R}{\partial z} + \frac{s}{v(\chi)}\,R = -\gamma_0 R + \frac{d}{M_c^2}\frac{1}{\chi}\frac{\partial}{\partial\chi}\left(\chi^{1-2q}\frac{\partial R}{\partial\chi}\right) \tag{1.185}$$

The moments σ_n of the distribution R may be obtained from Eq. (1.115) by replacing $\omega \to s$ and $P \to R$.

This equation is reminiscent of Eq. (1.174) and suggests that the solution for R is a linear combination of the steady-state solution already found in Eq. (1.175),

$$R(\chi,z,s) = \sum_{j=1}^\infty a_j(z,s)P_j(\chi)\,e^{-(\gamma_0+\gamma_j+sn/c)z} \quad . \tag{1.186}$$

The exponential factor, sn/c, is arbitrarily inserted to null out the delay common to all modes. When Eq. (1.186) is inserted into Eq. (1.185) the differential equation for the coefficients a_j is obtained:

$$\frac{\partial a_j}{\partial z} + s \sum_{j=1}^{\infty} M_{jk} a_j e^{-(\gamma_j - \gamma_k)z} = 0 \quad , \tag{1.187}$$

where,

$$M_{jk} = \int_0^1 2\chi\left(\frac{1}{v(\chi)} - \frac{n}{c}\right) P_j(\chi) P_k(\chi) d\chi \quad . \tag{1.188}$$

From the value of $v(\chi)$ Eq. (1.131) the matrix M_{jk}, to first order in Δ, is simply,

$$M_{jk} = \frac{n\Delta}{c}\left(\frac{\alpha - 2 - \varepsilon}{\alpha + 2}\right) \int_0^1 2\chi^{3\alpha + 2/\alpha + 2} P_j(\chi) P_k(\chi) d\chi \quad . \tag{1.189}$$

By using an iterative perturbation technique to solve Eq. (1.187), the pulse moments are found in terms of the quantity M_{jk}. The first three have been evaluated (1.77) and we examine the results.

The zeroth moment, σ_0, is the total power passing the point z. In the limit that $z > 1/\gamma_1$, it simply becomes the steady-state distribution already given in Eq. (1.182).

The first moment or mean delay time in this limit is

$$\tau(z) \cong \left[\frac{n}{c} + M_{11}\right] z + C_\tau \quad , \tag{1.190}$$

where C_τ is a constant depending on differences in the initial power distribution. As expected, the mean arrival increases linearly with z. The quantity M_{11} determines the effective velocity of the mode-coupled distribution.

Finally the asymptotic $(z > 1/\gamma_1)$ form of the rms pulse width is

$$\sigma(z) = \sqrt{2z}\left[\sum_{j=2}^{\infty} \frac{|M_{j1}|^2}{\gamma_j - \gamma_1}\right]^{1/2} \quad . \tag{1.191}$$

    This equation demonstrates the beneficial effect of
mode coupling on the information carrying capacity of a waveguide. Without coupling, the rms width increases directly
with guide length at a rate given by Eqs. (1.133) and (1.134).
However, with coupling, the rms width increases proportionally
to $(length)^{\frac{1}{2}}$, and hence the rate of decrease of information
capacity with length is considerably less.

    From the form of M_{jk} in Eq. (1.189), it is seen that
to first order in Δ, $\sigma(z)$ in the presence of mode coupling
exhibits the same minimum at $\alpha = 2 + \varepsilon$ as obtained for the
optimal gradient in Eq. (1.133) but now with a considerably
different length dependence.

1.5.3.2   Mode Coupling Attenuation

    We next consider the loss associated with this mode
coupling. To do this, as shown by Eqs. (1.167) and (1.172)
the curvature spectrum must be known. In general, this is not
possible. However, Olshansky (1.82) has found that for the
microbending case of a fiber passing over a large number of
bumps as in Figure 1.19 and subject to an elastic restoring
force, the curvature power spectrum can be generalized to

$$\langle C\,|(\Delta\beta)|^{2}\rangle = C_{o}(\Delta\beta D)^{-2\eta} \quad , \tag{1.192}$$

where C_0 characterizes the strength of the coupling, D measures
the correlation length of the curvature, and $\eta = 0$ corresponds
to a constant curvature power spectrum. Combining Eqs. (1.167),
(1.169), (1.172), (1.173), and (1.192) and substituting them
into Eq. (1.177) gives the excess loss caused by mode coupling
produced by this power spectrum,

$$\gamma_1 = \frac{C(\alpha,\eta)}{\Delta}\left(\frac{a^2}{\Delta}\right)^{\eta} \tag{1.193}$$

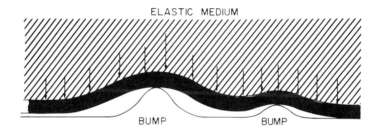

ELASTIC MEDIUM

BUMP   BUMP

Figure 1.19.   Schematic diagram of fiber configuration for
                which the distortion power spectrum can be
                calculated.

where

$$C(\alpha,\eta) = \frac{C_o}{2} \; \frac{1}{D^{2\eta}} \left(\frac{\alpha + 2}{4\alpha}\right)^{1+\eta} (1 + q)^2 z_1^2(\ell) \left(\frac{m}{M_c}\right)^{2(1+q)} \qquad \cdot \quad (1.194)$$

        The dependence of C on the index gradient is contained
in Eq. (1.194) and is plotted in Figure 1.20 for three values of
η. For $\eta = 2$ there is a two-fold increase in C in going from
a step ($\alpha = \infty$) to a parabolic ($\alpha = 2$) guide, while for $\eta = 0$
little difference in C between the two guide types exists.

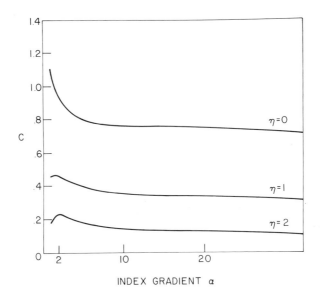

INDEX GRADIENT α

Figure 1.20.   Plot of the steady-state mode-coupling attenuation
              coefficient as a function of index gradient α for
              three different types of distortion.   [From
              Olshansky (1.77.)   Reproduced with permission.]

        The behavior of the loss with respect to the core
radius and the relative index difference bears examination.   It
is noted that the loss resulting from random curvatures decreases
as Δ increases and as the core radius decreases.   This would
suggest easy remedies for these type losses.   However, tradeoffs
obviously exist since as Δ increases the bandwidth decreases
(Eq. (1.133)), and as the core radius decreases the coupling
efficiency is reduced.   A careful balance for a particular
situation is, therefore, required.

1.5.3.3   Mode Coupling Pulse Propagation

A single parameter can be used to characterize mode coupling pulse characteristics quite well. From Eq. (1.191) the rms mode coupled pulse width can be put in the form,

$$\sigma(z) = \sigma_c(\alpha,\eta)\sqrt{\frac{z}{\gamma_1}} \quad ,$$
(1.195)

where,

$$\sigma_c = \left[2z_1^2 \sum_{\ell=2}^{\infty} \frac{|M_{1\ell}|^2}{z_\ell^2 - z_1^2} \right]^{1/2} \quad .$$
(1.196)

The length dependent behavior of $\sigma(z)$ is shown for both step ($\alpha = \infty$) and parabolic ($\alpha = 2$) waveguides in Figure 1.21. The guides are assumed to have $\Delta = 0.01$, a distortion spectrum $\eta = 0$, and excess mode coupling losses of 2 and 10 dB/km. It is seen that depending upon the excess loss, $\sigma(z)$ makes a smooth transition between two asymptotic forms,

$$\sigma(z) = \sigma_u z \qquad\qquad z \ll 1/\gamma_1$$

$$\sigma(z) = \sigma_c\sqrt{\frac{z}{\gamma_1}} \qquad\qquad z \gg 1/\gamma_1$$
(1.197)

where σ_u is obtained from Eqs. (1.133) and (1.134) for the two profile parameter values. The intersection between these two regions is called the coupling length, L_c,

$$L_c = \frac{1}{\gamma_1}\left(\frac{\sigma_c}{\sigma_u}\right)^2 \quad .$$
(1.198)

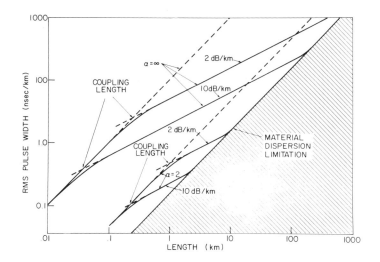

Figure 1.21.   RMS pulse width as a function of length for two
               different levels of mode coupling attenuation.
               [From Olshansky (1.77). Reproduced with per-
               mission.]

The total excess loss in the case of mode coupling is

$$\beta(dB) = 4.34 \ \gamma_1 \ z \quad . \tag{1.199}$$

If a quantity R is constructed, which is the ratio
of the mode coupled to the uncoupled rms pulse width, then it
is seen that

$$R^2 \beta = 4.34 \left(\frac{\sigma_c}{\sigma_u} \right)^2 = 4.34 \ \gamma_1 \ L_c = \text{constant} \quad . \tag{1.200}$$

Thus $R^2\beta$ is recognized as the mode coupling loss per unit coupling length. It depends only on α, η, Δ and on the initial modal power distribution, but it is independent of the fiber length. Because of this it gives a good measure of the tradeoff between the increased bandwidth and the decreased transmission caused by mode coupling. This quantity is plotted in Figure 1.22 as a function of α for a series of η values. It exhibits a value ~0.5 for the step guide, $\alpha = \infty$. The discontinuity near $\alpha = 2$ results because the sharp dip in $\sigma(z)$ occurs at slightly different α values with and without mode coupling. Therefore, R experiences a rather large change. Since practical waveguides may not have exact α-profiles, the near parabolic guide is expected to have $R^2\beta \sim 2$. Thus to achieve the same relative bandwidth increase for the near parabolic guide, a loss penalty ~4x greater than for the step guide will be incurred. This is shown in Figure 1.21 by the fact that for the same excess loss rate the coupling length for the step guide is one-quarter that for the parabolic waveguide. It is further noted in Figure 1.21 that the beneficial effects of mode coupling on the rms pulse broadening act only upon intermodel dispersion and do not overcome the fundamental dispersion limit of the material.

1.5.4   Protective Buffering

As the final consideration in the area of mode coupling, fiber buffering is considered briefly. In the process of packaging waveguides for practical applications, great care must be used to avoid incurring losses caused by random microbending (as just discussed). It has been shown (1.82), (1.83) that by sheathing the waveguide in a low modulus material, this loss can be reduced. Considering the model shown in Figure 1.19 and using the theory for random bending, the loss has been evaluated to be (1.82)

$$\gamma = N \langle h^2 \rangle \frac{a^4}{b^6 \Delta^3} \left(\frac{E}{E_f}\right)^{3/2} , \qquad (1.201)$$

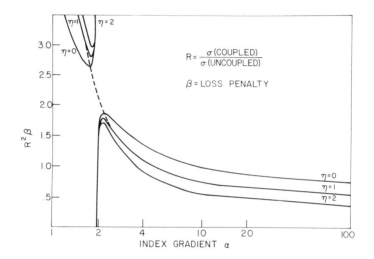

Figure 1.22.   Mode coupling parameter, $R^2\beta$ as a function index
               gradient for three types of distortion. [From
               Olshansky, (1.82). Reproduced with permission.]

where N is the number of bumps of average height h per unit
length, b is the total fiber diameter, a is the core radius,
and E_f and E are the elastic moduli of the fiber and the fiber
surrounding, respectively.

Thus it is seen that since E_f can be three orders of
magnitude larger than E for many plastic materials, this type
of shielding provides a means of reducing losses resulting from
mode coupling. Additionally, this loss decreases strongly with
increasing fiber size, which furnishes another design parameter.

1.6   SUMMARY

1.6.1   Design Considerations

          The optical propagation characteristics of the opti-
cal fiber discussed so far suggest many tradeoffs for enhancing
performance and minimizing manufacturing costs. Although an
optimized system design is complex and each system is somewhat
different, some standardization is possible. In an attempt to
summarize the effects that the key fiber parameters have on the
functional system performance, the tradeoff matrix in Table 1.2
is presented. The dependencies have all been discussed in this
chapter, but in most cases the values of the coefficients depend
on specific processes and material considerations. General
factors and directions of the effects are indicated. The total
system cost for each matrix element is, of course, the final
ajudicator of optimization. Some attempts have been made to
systematically obtain the indicated performance optimization for
10 km, 50 to 100 MBit/sec (1.84) and ~1 km, 20 MBit/sec systems
(1.85). For the first application, the values of the key fiber
parameters are Δ = 0.01, a core diameter of 50 μm, and a total
fiber diameter of 125 μm; for the second, the values are Δ = 0.02,
a core diameter of 100 μm, and a total diameter of 140 = μm. It
is expected that with time other systems will necessitate other
values for the key variables much as is the case with present
electrical conductors.

1.6.2   System Performance

          As a further summary of the considerations presented
in this chapter, the possible fiber bit rate as a function of
fiber length may be plotted, as in Figure 1.23.

Table 1.2. Dependence of the waveguide system optical performance upon the four primary waveguide design parameters and their variation. The interrelationship between various aspects of system performance and these parameters lead to design tradeoffs.

PARAMETERS WAVEGUIDE	SYSTEM ATTENUATION			INPUT COUPLING	BANDWIDTH
	Intrinsic	External Perturbation	Interconnection		
Profile α $\delta\alpha$	$\dfrac{\delta\alpha}{\alpha}$		$\dfrac{\delta\alpha}{\alpha}$	$\dfrac{\alpha}{\alpha+2}$	$[\alpha-\alpha_o(\lambda)]^{-1}$
Numerical Aperture NA δNA	NA^2 $\dfrac{\delta NA}{NA}$	$NA^{-2(n+1)}$	$\dfrac{\delta NA}{NA}$	NA^2	NA^{-2}
Core Diameter a δa	$\dfrac{\delta a}{a}$	a^2	$\dfrac{\delta a}{a}$	a^2	
Cladding Diameter b δb	$e^{-\kappa(b-2a)}$ $\dfrac{\delta b}{b}$	b^{-2n}			

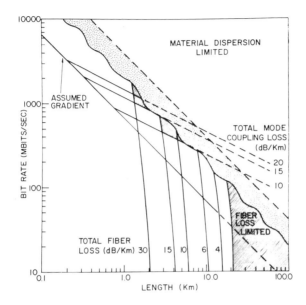

Figure 1.23.   Transmission bit rate as a function of fiber
                length.

          A GaAs laser operating a 900 nm with 1 mwatt coupled
into the fiber, a $\Delta = 0.01$, a Si-APD detector operating at
optimum gain, an error rate of 10^{-9}, an $R^2\beta = 0.5$, and a 5%
deviation from the optimal index gradient is assumed. Material
dispersion bounds the plot on the upper right. At some length,
loss in the waveguide will produce a signal-limited condition
as shown by the near vertical lines. Within the remaining
region, systems may be considered. As shown, mode coupling can
enhance the transmission bandwidth, being traded off, however,
against having to decrease the overall fiber loss or the system
length. A similar plot could be done for longer operating wavelengths where the material dispersion and attenuation-imposed
boundaries would be shifted to the right.

An attempt has been made to cover the operation of optical fiber waveguides in some detail. The reader is referred to the original works for more complete information. At present, the future looks bright for a vast variety of systems to use the unique capabilities of optical waveguides. General concepts of the type presented here will provide the base upon which future progress will be built.

PROBLEMS FOR CHAPTER 1

1. Using the ray model for a step waveguide, show that at
    each reflection the axial angle, Θ, is conserved for an
    arbitrary ray. In the modal representation, this corre-
    sponds to the modal propagation constant being preserved
    along the waveguide. (Hint: see Eq. 1.5.)

2. Derive the ray capture equation (1.8) starting with the
    vector relation for total internal reflection (1.7).

3. Show that for a monotonically decreasing radial refrac-
    tive index distribution the ray path is sinusoidal.

4. Show that in the paraxial approximation (first order in
    α) the "square law," "the hyperbolic secant," and the
    "inverse square law" index distributions are identical.

5. Calculate the "focal length" of a square law graded
    medium with $\Delta = 0.01$ and a = 1 mm; with 0.05 and 5 mm.

6. Sketch the transverse electric field power distribution
    of the EH_{11}, TE_{01}, TM_{01}, and HE_{21} modes. Sketch the
    admixture of the TE_{01}, TM_{01}, and HE_{21} modes.

7. Calculate the radius within which 50% of the HE_{11} mode
    power is contained; 80% radius, 95% radius.

8. Derive expressions in terms of the mode group number for
    the inner and outer turning points r_1 and r_2 of the step
    profile and parabolic waveguides.

9. Show that the near-field intensity distribution for leaky
    modes in an α-profile waveguide assuming a uniform mode
    excitation is given by:

$$\frac{1}{2\pi R} \frac{dP}{dR} = \pi a^2 \sin^2 \theta_c (1 - R^\alpha) \left[\frac{1}{\sqrt{1 - R^2}} - 1 \right] \; .$$

10. Show that the far-field intensity distribution for leaky
    modes in a waveguide with $\alpha = 2$ index profile assuming
    uniform excitation is

$$\frac{1}{2\pi\eta} \frac{dP}{d\eta} = \pi a^2 \sin^2 \theta_c (1 - \eta^2)^{\frac{1}{2}} \qquad (\eta \leq 1) \; .$$

11. The bandwidth of a perfect waveguide at 500 nm is 200
    MHz-km and the guide has a maximum bandwidth at 1100 nm
    of 20 GHz-km. A perturbed waveguide with the same compo-
    sition and peak wavelength has a bandwidth of only 1500
    MHz-km at 1100 nm. What is its bandwidth at 500 nm?
    Assume the same guide is measured to be 150 MHz at
    500 nm. What is the bandwidth at 1100 nm?

12. The transmitted pulse shape of a waveguide is $p(t) = t^n \exp(-\alpha t)$ for $t > 0$ and $p(t) = 0$ for $t < 0$.
    What is the -3 dB frequency? The -10 dB frequency?

13. Calculate the rms width of the gaussian pulse,

$$p(t) = e^{-(\alpha t)^2} \; .$$

    Compare this with the rms width of the pulse in Problem 12.

14. Derive the relationship between the full width of half maximum (FWHM) and the rms width of a gaussian pulse; a "square" pulse; a "triangle" pulse.

15. For what bend radius in a waveguide with $\Delta = 0.01$ and $a = 25$ μm are 10% of the modes lost? 50% lost? Assume $\Delta = 0.02$ and $a = 50$ μm. What is the change in bend radius for the same losses?

16. The elastic modulus of silica is 1×10^7 psi and for silicone ~500 psi. Assuming $\Delta = 0.01$, a and b of 25 μm and 62 μm, respectively, and 100 bumps/kilometer, what average bump height is required to produce a loss of 10 dB/km?

17. Assuming a parabolic waveguide with 4 dB/km excess loss caused by perturbations, what is the coupling length?

18. A step profile waveguide with $\Delta = 0.01$ and $N = 1.477$ has 10 dB/km excess loss as a result of perturbations. To first order in Δ, and ignoring the effect of profile dispersion, for what length will its pulse broadening be equal to that of a graded index waveguide with a profile, $\alpha = 2.1$?

REFERENCES

1.1   J. Tyndell, "On Some Phenomena Connected with the Motion of Liquids," *Proc. Royal Institution*, *1*, 446   (1854).

1.2   A. G. Bell, *Proc. Am. Assoc. Advancement Sci.*, *29*, 115 (1880).

1.3   P. Kaiser, E. A. J. Marcatili, and S. E. Miller, *Bell Syst. Tech. J.*, *52*, 265   (1973).

1.4   S. Kawakami and S. Nishida, *J. Quantum Electron*, *QE-10* (1974).

1.5   H. Kita, I. Kitano, T. Uchida, and M. Furukawa, "Light-Focussing Glass Fibers and Rods," *J. Amer. Ceram. Soc.*, *54*, 321   (1970).

1.6   E. G. Rawson, D. R. Herriott, and J. McKenna, "Analysis of Refractive Index Distributions in Cylindrical, Graded-Index Glass Rods (GRIN Rods) Used as Image Relays," *Appl. Opt.*, *9*, 753   (1970).

1.7   P. J. Sands, *J. Opt. Soc. Am.*, *60*, 1436, (1970).

1.8   W. Streifer and K. B. Paxton, *Appl. Opt.*, *10*, 769 and 1164   (1971).

1.9   D. T. Moore, *J. Opt. Soc. Am.*, *61*, 886, (1971).

1.10   E. W. Marchand, *Gradient Index Optics*   (Academic Press, New York, 1978).

1.11   J. A. Stratton, *Electromagnetic Theory*   (McGraw-Hill Book Company, New York, 1941).

1.12   E. Snitzer, "Cylindrical Dielectric Waveguide Modes," *J. Opt. Soc. Am.*, *51*, 491   (1961).

1.13   D. Marcuse, *Light Transmission Optics*   (Van Nostrand Reinhold Company, New York, 1972).

1.14   D. Marcuse, *Theory of Dielectric Waveguides*   (Academic Press, New York, 1974).

1.15   F. P. Kapron, N. F. Borrelli, and D. B. Keck, *IEEE J. Quantum Electron.*, *QE-8*, 222   (1972).

1.16   I. P. Kaminow and V. Ramaswamy, *Appl. Phys. Lett.*, *34*, 268   (1979).

1.17   G. Sagnac, *Comptes Rendus*, *157*, 708   (1913).

1.18   A. W. Snyder, "Asymptotic Expressions for Eigenfunctions and Eigenvalues of a Dielectric Optical Waveguide," *IEEE Trans. Microwave Theory Tech.*, *MTT-17*, 1310   (1969).

1.19   D. Gloge, "Weakly Guiding Fibers," *Appl. Opt.*, *10*, 2252 (1971).

1.20   E. Merzbacker, *Quantum Mechanics*   (John Wiley and Sons, New York, 1961), Chap. 7.

1.21   C. N. Kurtz and W. Streifer, "Guided Waves in Inhomogeneous Focusing Media," *IEEE Trans. Microwave Theory Tech.*, *MTT-17*, 250   (1969).

1.22   D. Gloge and E. A. J. Marcatili, "Multimode Theory of Graded Core Fibers," *Bell Syst. Tech. J.*, *52*, 1563 (1973).

1.23   A. W. Snyder and D. J. Mitchell, *Electron. Lett.*, *9*, 437 (1973).

1.24   M. J. Adams, D. N. Payne, and F. M. E. Sladen, *Electron. Lett.*, *11*, 238   (1975).

1.25   D. B. Keck, "Spatial and Temporal Power Transfer Measurements on a Low Loss Optical Waveguide," *Appl. Opt.*, *13*, 1882   (1974).

1.26   R. Olshansky, S. M. Oaks, and D. B. Keck, "Topical Meeting on Optical Fiber Transmission II," Williamsburg, Virginia (1977).

1.27   R. Olshansky and S. M. Oaks, *Appl. Opt.*, *17*, 1830   (1978).

1.28   D. B. Keck, R. D. Maurer, and P. C. Schultz, "On the Ultimate Lower Limit of Attenuation in Glass Optical Waveguides," *Appl. Phys. Lett.*, *22*, 307   (1973).

1.29   K. J. Beales, C. R. Day, W. J. Duncan, A. G. Dunn, P. L. Dunn, and G. R. Newns, "Low Loss Graded Index Fibers by the Double Crucible Technique," *Physics Chem. Glasses*, *21*, 25   (1980).

1.30   D. S. Carson and R. D. Maurer, "Optical Attenuation in Titania Silica Glasses," *J. Non-Cryst. Solids*, *11*, 268 (1973).

1.31   P. Kaiser, *J. Opt. Soc. Am.*, *64*, 475 (1974).

1.32   R. D. Maurer, E. J. Schiel, S. Kronenberg, and R. A. Lux,
       "Effect of Neutron and Gamma-Radiation on Glass Optical
       Waveguides," *Appl. Opt.*, *12*, 2024 (1973).

1.33   E. J. Friebele, M. E. Gingerich, and G. H. Sigel, *Appl.
       Phys. Lett.*, *32*, 619 (1978).

1.34   E. J. Friebele, P. C. Schultz, M. E. Gingerich, and
       L. M. Hayden, "Topical Meeting on Optical Fiber Trans-
       mission," *Paper TUG1*, Washington, D.C. (March, 1979).

1.35   I. D. Aggarwal, P. B. Macedo, and C. J. Montrose, "Light
       Scattering in Lithium Aluminosilicate Glass System,"
       American Ceramic Society Meeting, Bedford Springs,
       Pennsylvania (1974).

1.36   P. B. O'Conner, J. B. MacChesney, and F. V. D. Marcello,
       Second European Conf. Opt. Fiber Comm., *Paper II. 3*,
       Paris (1976).

1.37   R. H. Stolen, E. P. Ippen, and A. R. Tynes, "Raman
       Oscillation in Glass Optical Waveguides," *Appl. Phys.
       Lett.*, *20*, 62 (1972).

1.38   R. G. Smith, "Optical Power Handling Capacity of Low Loss
       Optical Fibers as Determined by Stimulated Raman and
       Brillouin Scattering," *Appl. Opt.*, *11*, 2489 (1972).

1.39   J. D. Crow, "Power Handling Capability of Glass Fiber
       Lightguides," *Appl. Opt.*, *13*, 467 (1974).

1.40   S. M. Jensen and M. K. Barnoski, "Topical Meeting on
       Optical Fiber Transmission II," *Paper TuD7*, Williamsburg,
       Virginia (1977).

1.41   L. G. Cohen and C. Lin, *Appl. Opt.*, *16*, 3136 (1977).

1.42   D. Gloge, "Propagation Effects in Optical Fibers," *IEEE
       Trans. Microwave Theory Tech.*, *MTT-23*, 106 (1975).

1.43   A. H. Cherin and E. J. Murphy, *Bell System Tech. J.*, *54*
       (1975).

1.44   C. Pask, A. W. Snyder, and D. J. Mitchell, "Number of
       Modes on Optical Waveguides," *J. Opt. Soc. Am.*, *65*, 356
       (1975).

1.45   S. D. Personick, "Receiver Design for Digital Fiber Optic
       Communication Systems," *Bell Syst. Tech. J.*, *52*, 843
       (1973).

1.46   F. P. Kapron and D. B. Keck, *Appl. Opt.*, *10*, 1519   (1971).

1.47   D. Gloge, "Dispersion in Weakly Guiding Fibers," *Appl.
       Opt.*, *10*, 2442   (1971).

1.48   L. G. Cohen, C. Lin, and W. G. French, *Electron. Lett.*,
       15, 334   (1979).

1.49   R. Olshansky and D. B. Keck, *Appl. Opt.*, *15*, 483   (1976).

1.50   I. H. Malitson, *J. Opt. Soc. Am.*, *55*, 1205   (1965).

1.51   I. P. Kaminow and H. M. Presby, *Appl. Opt.*, *15*, 3029
       (1976).

1.52   D. N. Payne and A. H. Hartog, *Electron. Lett.*, *13*, 627
       (1977).

1.53   F. M. E. Sladen, D. N. Payne, and M. J. Adams, *Electron.
       Lett.*, *15*, 469   (1977).

1.54   R. Olshansky, *Appl. Opt.*, *15*, 782   (1976).

1.55   J. A. Arnaud and W. Mammel, *Electron. Lett.*, *12*, 7
       (1976).

1.56   A. W. Snyder and R. A. Sammut, *Electron. Lett.*, *15*, 269
       (1979).

1.57   A. W. Gambling, H. Matsumura, and C. M. Ragsdale, *Electron.
       Lett.*, *15*, 475   (1979).

1.58   M. Eve, K. Hartog, R. Kashyap, and D. N. Payne, Fourth
       European Conference on Optical Fiber Communications,
       Paper II.2, Genoa   (1978).

1.59   D. B. Keck and R. Bouillie, *Opt. Commun.*, *25*, 43   (1978).

1.60   I. P. Kaminow and H. M. Presby, *Appl. Opt.*, *16*, 108
       (1977).

1.61   E. A. J. Marcatili, *Bell System Tech. J.*, *56*, 49   (1977).

1.62   R. Olshansky, *Electron. Lett.*, *14*, 330 (1978).

1.63   S. Geckler, *Electron. Lett.*, *15*, 682 (1979).

1.64   A. Weirholt, *Electron. Lett.*, *15*, 733 (1979).

1.65   M. G. Blankenship, D. B. Keck, P. S. Levin, W. F. Love,
       R. Olshansky, A. Sarkar, P. C. Schultz, K. D. Sheth, and
       R. W. Siegfried, Topical Meeting on Optical Fiber Commun-
       ication, *Paper PD-3*, Washington, D.C. (1979).

1.66   W. F. Love, Sixth European Conference on Optical Fiber
       Communications, York, England (1980).

1.67   M. Eve, *Electron. Lett.*, *13*, 315 (1977).

1.68   M. Eve, *Opt. Quant. Electron.*, *10*, 41 (1978).

1.69   E. A. J. Marcatili and S. E. Miller, "Improved Relations
       Describing Directional Control in Electromagnetic Wave
       Guidance," *Bell Syst. Tech. J.*, *48*, 2161 (1969).

1.70   F. P. Kapron, D. B. Keck, and R. D. Maurer, "Radiation
       Losses in Glass Optical Waveguides," *Appl. Phys. Lett.*,
       17, 423 (1970).

1.71   D. Gloge, *Appl. Opt.*, *11*, 2506 (1972).

1.72   D. Marcuse, *J. Opt. Soc. Am.*, *66*, 216 (1976).

1.73   A. W. Gambling and H. Matsumura, *Electron. Lett.*, *13*, 532
       (1977).

1.74   H. Heiblum and J. H. Harris, *IEEE J. Quantum Electron.*,
       QE-1, 75 (1975).

1.75   R. Olshansky and D. A. Nolan, *Appl. Opt.*, *15*, 1045 (1976).

1.76   D. Gloge, "Optical Power Flow in Multimode Fibers," *Bell
       Syst. Tech. J.*, *51*, 1767 (1972).

1.77   R. Olshansky, "Mode Coupling Effects in Graded-Index
       Optical Fibers," *Appl. Opt.*, *14*, 935 (1975).

1.78   L. Jeunhomme and J. P. Pochelle, *Electron. Lett.*, *11*, 425
       (1975).

1.79   K. Furuya and Y. Suematsu, *Appl. Opt.*, *19*, 1493   (1980).

1.80   K. Peterman, *Electron. Lett.*, *12*, 107   (1976).

1.81   R. Olshansky, Second European Conference on Optical Fiber
       Communications, Paris   (1976).

1.82   R. Olshansky, "Model of Distortion Losses in Cabled
       Optical Fibers," *Appl. Opt.*, *14*, 20   (1975).

1.83   W. B. Gardner and D. Gloge, "Microbending Loss in Coated
       and Uncoated Optical Fibers," Topical Meeting on Optical
       Fiber Transmission, Williamsburg, Virginia (1975).

1.84   H. Murata, S. Ingo, Y. Matsuda, and T. Kuroha, Fourth
       European Conference on Optical Fiber Communications,
       Paper IV.7, Genoa   (1978).

1.85   A. J. Morrow, R. G. Sommer, D. B. Keck, and L. C. Gunderson,
       "Design and Fabrication of Optical Fiber for Data
       Processing Applications," Topical Meeting on Optical Fiber
       Communication, *Paper WF2*, Washington, D. C.   (1979).

CHAPTER 2

OPTICAL FIBER CABLE

James E. Goell

Lightwave Technologies Inc.,
Los Angeles, California 91406

2.1   INTRODUCTION

Today, optical fiber cables are being installed routinely
in a large variety of environments. Cables have been successfully installed in ducts, placed directly in the ground, and
lashed to telephone poles. Furthermore, cables are now being
developed which should survive the rigors of battlefield use and
installation from helicopters.

Routine cable installation is possible because of the
success of a concerted effort by numerous organizations to overcome problems with cabling-induced attenuation and fiber breakage.
As drawn, optical fibers are extremely strong; however, abrasion
and water attack seriously reduce the strength of unprotected
fibers. Furthermore, improper cabling can introduce bends which
cause optical radiation and thus increase optical attenuation.
Plastic coating techniques have been developed to preserve fiber
strength and to inhibit cabling-induced attenuation. In parallel,
cabling designs and techniques have been developed that allow
plastic-coated fibers to be incorporated into fieldable cables
with little or no cabling-induced attenuation.

This chapter describes optical cable fundamentals
governing mechanical and optical performance and life, and the
state of the art of cable performance. It should provide the background needed for system designers to make cable selections, serve
as an introduction for future cable designers to basic design considerations, and form a basis for fiber designers to make design
tradeoffs.

2.2   OPTICAL CABLE FUNDAMENTALS

Both otpical and mechanical performance over time must
be considered in designing and using optical cables. These areas
are addressed in the following sections.

2.2.1   Optical Performance

Cable insertion loss is comprised of the sum of the
attenuation (length dependent) and short length (length independent) losses. Both sources of loss must be considered when selecting or designing a cable for a specific application.

2.2.1.1 Attenuation

Cable attenuation is the sum of the attenuation of the
uncabled fiber and the attenuation induced in the fiber by
cabling.

Uncabled fiber attenuation is discussed in detail in
Chapter 1. Therefore, it will not be discussed further here
except to point out that it is not completely independent from
cabling-induced attenuation. The susceptibility of a fiber to
cabling-induced attenuation decreases as numerical aperture
increases. (This effect is discussed further in a subsequent
section.) However, numerical aperture is usually increased by
increasing core doping. Increased core doping causes Rayleigh
scattering, a component of intrinsic attenuation, to increase.

Cabling-induced attenuation is caused by radiation from small-radius fiber bends. These bends are caused by transverse forces applied to the fiber at the location of an internal bump in the cable. Compressive longitudinal force also can cause fiber buckling.

Figure 2.1 shows fiber subjected to transverse stress over a bump in a cable. According to Olshansky (2.1), the dependence of the distortion loss on the core radius r_c and the relative index difference Δ is given for a step index fiber by

$$\alpha_b = \frac{cr_c^4}{\Delta^3} , \qquad\qquad (2.1)$$

where

$$c = 0.9\ p\ \frac{h^2}{b^6} \left(\frac{E_m}{E_s}\right)^{3/2}$$

α_b = distortion loss

r_c = core radius

b = fiber diameter

Δ = relative index difference $1 - \left(\dfrac{n_{clad}}{n_{core}}\right)$

h = effective rms bump height

p = number of bumps per unit length

E_m = modulus of encapsulating material

E_s = modulus of fiber core

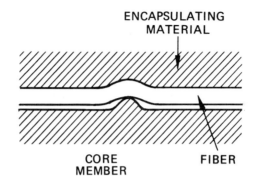

ENCAPSULATING
MATERIAL

CORE   FIBER
MEMBER

Figure 2.1.   Transverse stress induced bend.

With the assumed values

r_c = 50 μm

b = 125 μm

Δ = 0.0053 (numerical aperture = 0.15)

E_m = 690 N/mm^2

E_s = 62,000 N/mm^2

the loss would be 0.0007 dB for each 1.0 μm bump.

Examination of Eq. (2.1) shows that excess cable losses
drop rapidly with increasing fiber diameter and numerical aperture
and with decreasing core-cladding ratio. This equation does not
strictly apply for graded index and single-mode fibers. However,
it can be expected to be qualitatively correct.

A plastic coating can stiffen a fiber and provide cushion-
ing against bumps, thus reducing microbend loss. In fact, the
use of multiple coatings is now common. It has been shown by
Gloge (2.2) that such coatings more effectively inhibit microbend
loss than single layer coatings. Figure 2.2 shows Gloge's computed results for the following conditions:

Inner jacket modulus	10 N/mm^2
Outer jacket modulus	$1,000 \text{ N/mm}^2$
Fiber radius	0.06 mm
First jacket radius (soft outer jacket)	0.02 mm
First jacket radius (hard outer jacket)	0.04 mm

For these conditions, the curve shows that for a jacket radius greater than 0.25 mm, double jackets are significantly superior. The superiority of the hard outer shell over the soft outer shell, in general, is yet to be proved. The difference between the two approaches shown in Figure 2.2 is small and may vary drastically with the specific conditions. Nevertheless, the hard-shell approach has proved to be practical and is widely employed.

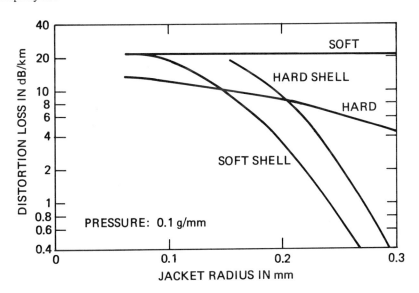

Figure 2.2. Distortion loss versus outside-jacket radius. (Key: Soft — single soft jacket, hard — single hard jacket, hard shell — hard outer jacket/soft inner jacket, soft shell — soft outer jacket/ hard inner jacket.)

All cable manufacturers are now using jacketed fibers.
Figure 2.3 shows some of the approaches now in use. In some cases,
single-jacketed fibers are being employed in a loose cable construction. This approach gives low cabling-induced attenuation,
but does not minimize cable size. In other cases, the "filled"
loose tube and "tight" construction fibers are being used.

Low-level tension has a negligible effect on fiber attenuation so long as it does not induce bends. C. Kao and G. Bickel
have performed experiments to evaluate the effect of stress on
the attenuation of fibers. For all of their tests, they used
0.23 numerical aperture, 50 μm core diameters, 125 μm cladding
diameter Kynar-coated chemical vapor deposition (CVD) fibers. The
purpose of the tests was to simulate the effects of elongation,
tension over bumps, transverse stress-induced bending, and bends
induced by residual stresses in the fiber coating. The tests and
their results are listed below.

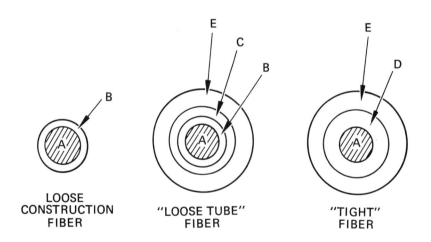

LOOSE
CONSTRUCTION   "LOOSE TUBE"   "TIGHT"
FIBER   FIBER   FIBER

Figure 2.3.   Fiber jacket configuration. (A — glass fiber, B —
              thin protective jacket, C — air or liquid, D — soft
              layer, and E — jacket.)

1. The fiber was wound on a drum and the drum
    was heated, producing a 0.2% increase in drum
    circumference. The effect of elongation, and
    tension-induced bends caused by fiber coating
    and drum irregularities were evaluated. No
    change in attenuation was observed over the
    full range of the test.

2. The fiber was muliply wound between two drums
    and the fibers strained to 0.2% by separating
    the drums to observe the same phenomenon as
    in test 1. Again, no attenuation change was
    observed.

3. A 0.254-mm-diameter rod was inserted under the
    fibers wound and strained as described in
    test 2 to observe tension-induced bend loss.
    Again, no attenuation change was observed.

4. The fibers wound as in test 2 were interwoven
    with three 0.254-mm-diameter rods to observe
    the effect of transverse stress-induced bend-
    ing on fiber loss. For this test, a signifi-
    cant increase was measured as shown in
    Figure 2.4.

2.2.1.2 Short Length Losses

Short-length losses consist of the sum of the Fresnel
(end reflection) loss at the source, the Fresnel loss at the
detector, the source collection loss, the detector collection loss,
and the connector and splice losses. The Fresnel loss is given
by

$$10 \log \frac{n_c - n_1}{n_c + n_1} , \qquad (2.2)$$

where n_c is the core refractive index and n_1 is the launching
medium refractive index. The source collection loss decreases
with increasing numerical aperture and source active area-fiber
core overlap. Detailed treatment of source collection loss is
given in Chapter 3.

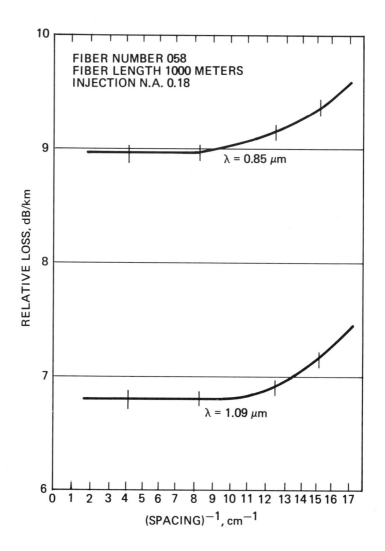

Figure 2.4.   Loss increase caused by induced bends.
(1 km of fiber wrapped around two 0.3 m
diameter drums 6 m apart, bends induced
by interweaving fiber with 3, 0.254 mm
rods separated as indicated on abscissa.)

2.2.1.3 Optical Tradeoffs

For a given system, a tradeoff must be made between short- and long-length losses. For very short cables, the end losses make the dominant contribution to the total insertion loss, whereas for very long cables the attenuation will dominate. Thus, for some short-distance applications with large-area sources, it is advantageous to use large numerical aperture, large-core fibers (or even a number of such fibers in parallel); for long cables, small core, lower numerical aperture fibers are best.

2.2.2 Mechanical Performance

The primary concerns of cable design are tensile strength, impact resistance, crush resistance, bending fatigue, and torque balance. Gas or water blockage is also a concern for some applications.

The nature of these concerns depends on the application. For example, for duct installation, long lengths of cable are subjected to tension during installation. Once installed, residual tensions are a small fraction of the installation load. On the other hand, direct-buried cables experience little tension either during installation or service, but they must resist crushing and possible shearing by rocks. Lashed aerial cables experience little tension or crushing forces. However, they are exposed to the weather extremes of the local environment. For battlefield use, installation tensions should be low since these cables are payed out on the ground. Impact and crush resistance must be high to withstand repeated vehicle crossings. Torque balance is important when cables are to be used under tension. This is to be expected for a variety of undersea applications where cables

are to be used vertically. Without torque balance, in these
applications cables can literally unwind. Finally, for undersea cables, water blockage is important or a cable break could
lead to equipment flooding. Similar, though less severe, problems
exist for buried cables.

Fiber tensile strength is an important cable design consideration. It can drive the selection of fabrication techniques
and force the incorporation of strength members which complicates
cable design and increases cable size. During cable production,
fibers must be wound onto spools and passed through cabling
machinery. Unless fibers have sufficient strength, these opera-
tions are uneconomical and impracticable.

Under typical conditions, strength members provide most
of the strength of cables; the fibers will undergo strain along
with the cable. It is fiber strain rather than strength that is
important under such conditions. However, it has become customary
to examine fiber stress. Since fiber stress is proportional to
strain, at least to levels well above those that are practical,
this inconsistency is not important.

It is possible to design cables so that the initial cable
strain resulting from tension does not cause fiber elongation.
But this design is achieved at the expense of an increase in cable
diameter. In general, a few tenths of a percent, at most, offset
in the cable stress-fiber strain curve can be achieved without a
major increase in cable size (2.3).

The design of a cable involves several tradeoffs.
Strength members are used to increase strength. However, it is
desirable to minimize the use of strength members to minimize
cable size. Furthermore, the more strengthening materials and
fiber incorporated into a cable, the stiffer it will be. To
overcome stiffness, helical (stranded) structures are often
employed. These structures bend the fibers. If the lay length

of the helix is too short, bend radiation and fiber surface stress
may be unacceptable; but if it is too long, the cable may be too
stiff. Thus, lay length must be selected carefully. The helical
lay also causes torque imbalance. If torque balance is required,
contrawound strength members can be used. Alternatively, in some
cases, braiding can be employed.

2.2.2.1 Fiber Strength

The tensile strength of pristine glass fiber is compar-
able to that of the strongest materials, including steel. However, mechanical flaws can greatly reduce fiber strength. Furthermore, mechanical flaws can grow under stress in the presence
of OH^- ions leading to strength degradation with time.

High strength fibers are produced by

● Employing flaw-free preforms

● Drawing in a clean environment

● Controlling draw temperature

● Using protective coating.

The most important consideration is the use of protective coating
to prevent surface flaws. A variety of plastic materials and
coating techniques exist that are applicable to the protection
of optical fibers. Major factors to be considered in their
selection are

● Protection against mechanical damage

● Protection against moisture

● Processability and compatibility with fiber
  manufacturing

● Compatibility with the jacketing process.

The most common coatings are Kynar, ethelene vinyl acetate
(EVA), silicon room temperature vulcanizing (RTV), epoxy acrylate,

and urethane acrylate. All are applied by dip coating the fiber
during the draw process, as shown in Figure 2.5. Kynar, EVA, and
silicone RTV are heat cured (for Kynar and EVA to drive off a sol-
vent, and for silicone RTV to initiate a chemical reaction). With
epoxy acrylate and urethane acrylate, ultraviolet light is used to
initiate a chemical reaction.

Fiber strength can only be determined by destructive
tests. Two common approaches are short-sample testing and proof
testing.

For short-sample testing, a large number of samples about
one meter long are stressed to destruction either by applying
tension or wrapping them around a small radius mandrel. Figure 2.6
shows a Weibull plot (probability of failure versus stress plotted
on a linearizing scale) for a high-strength fiber. Some fibers
have occasional weak spots. Such fibers typically exhibit a few
weak points to the left of the trend line at low probability.

Proof testing is performed by sequentially applying tension to the whole fiber. This approach is commonly used as part
of the draw process. Strength is evaluated by assessing the number of breaks at a given stress level. Fibers as long as 10 km
have been successfully proof tested at stress levels to $1,400 \text{ N/mm}^2$.

As previously mentioned, fiber flaws grow when the fiber
is stressed and OH^- ions are present. To assure long cable life,
these conditions must be considered. Three approaches to assuring
long fiber life are (1) stress derating, (2) hermetic sealing,
and (3) surface compression.

Statistical models of aging have been proposed to predict fiber life. However, extensive proof-test data are required
to apply these models. To date, adequate data on fibers produced
with sufficient control are unavailable. Fortunately, theory
and data exist which show that the time to failure for a given
flaw increases rapidly as stress decreases. The theoretical rela-
tion between times to failure, t_1 and t_2, applied stresses σ_1
and σ_2, is given by

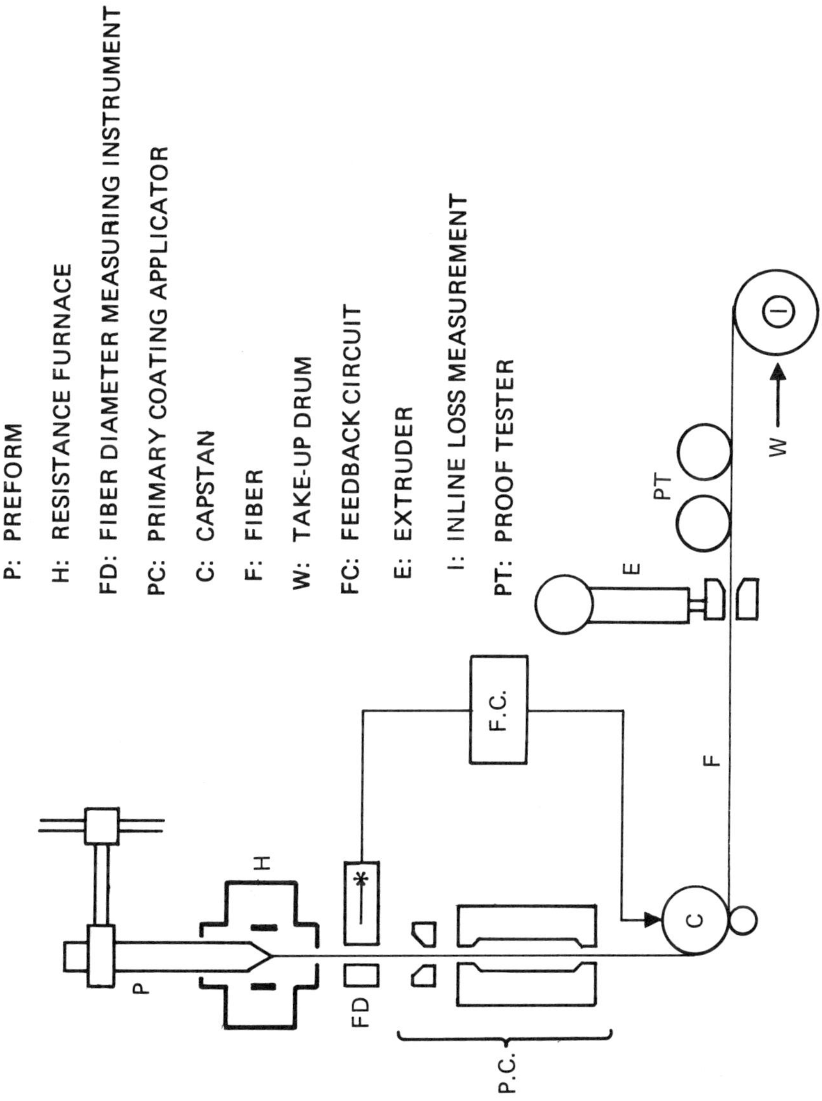
P: PREFORM
H: RESISTANCE FURNACE
FD: FIBER DIAMETER MEASURING INSTRUMENT
PC: PRIMARY COATING APPLICATOR
C: CAPSTAN
F: FIBER
W: TAKE-UP DRUM
FC: FEEDBACK CIRCUIT
E: EXTRUDER
I: INLINE LOSS MEASUREMENT
PT: PROOF TESTER
P
H
FD
P.C.
F.C.
C
F
E
PT
W
I

Figure 2.5. Fiber draw schematic.

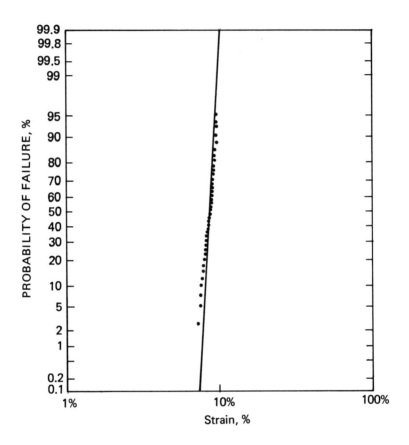

Figure 2.6   Weibull plot for a high strength fiber.

$$\frac{t_2}{t_1} = \log^{-1} N \log \frac{\sigma_1}{\sigma_2} . \qquad (2.3)$$

Figure 2.7 shows time to failure versus stress for fibers wrapped on a mandrel. For these fibers, which were measured in dry air, N was 24. Figure 2.8 shows similar data taken in water and air. When water was present, N was 15 compared to 24 in air. Numerous other tests have been run to determine the effect of moisture, temperature, and the acidity of the environment. It has been shown that N increases if the atmosphere is dry or acidic, as shown in Figure 2.9. Also, N decreases slowly if temperature is increased. In general, a derating from the proof stress level of three will give a high probability of a 50-year life.

Metal fiber coating has been successfully used to hermetically seal fibers (2.4). Such fibers exhibit dramatically improved durability. However, hermetically sealed fibers that can survive the required proof test levels have low reproducibility.

An excellent method for extending fiber life is to use a lower coefficient of expansion material for its surface than its core (2.5). When such a fiber cools after drawing, the fiber surface will be subjected to compressive stress. This stress counteracts the applied stress and, as a result, extends life. Fibers with 350 to 700 N/mm^2 surface compression have been built and may become commercially available.

2.2.2.2 Cabling-Induced Stress

When fibers are incorporated into a cable by wrapping them around a central strength member or around each other, stress is introduced. If the fibers are tightly wrapped around a core of strength members or other fibers, the fibers will be in tension. Such tension must be kept to a safe level.

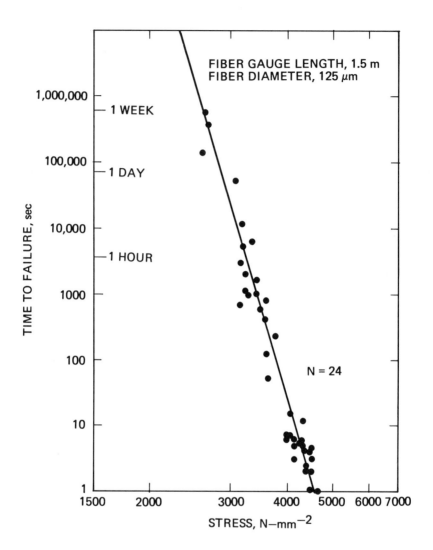

Figure 2.7.   Time to failure versus stress level — mandrel fatigue
            test.

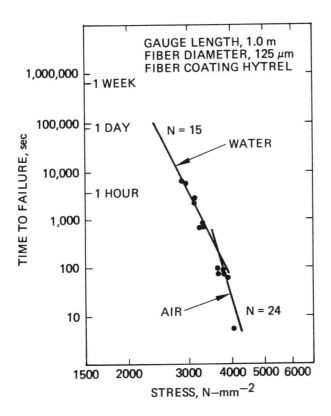

Figure 2.8.   Effect of water on time to failure versus stress.

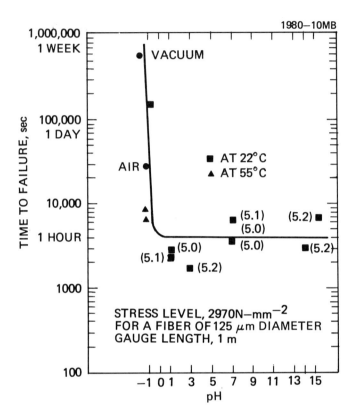

Figure 2.9.   Time to failure versus pH.

Wrapping the fibers around a core also induces torsion. Figure 2.10 shows a typical machine used to cable conventional conductors with low torsion. The large wheel of the machine rotates as the cable is drawn through it. For low lay angles (small pitch) this would introduce one rotation of the fiber for each wrap around the core. To compensate for twist to reduce conductor torsion, a series of gears and chains are incorporated behind the wheel so that as it rotates, the wire spool axes maintain a fixed orientation.

Figure 2.10.   Planetary stranding machine (Kabmatik).

When a fiber is bent around the strength member, flexural stress is induced. The peak flexural stress (peak surface stress) is given by

$$\sigma_f = \frac{Eb}{2R} \tag{2.4}$$

$$R = R_m \left[1 + \left(\frac{P}{2\pi R_m} \right)^2 \right] \tag{2.5}$$

where

R = fiber radius of curvature

E = modulus of elasticity

b = unjacketed fiber diameter

$R_m = \dfrac{b + D}{2} + t$

D = cable core diameter

t = fiber jacket thickness

p = pitch or lay.

The curves in Figure 2.11 show the fiber bend radius and flexural stress as a function of helix pitch for central strength membered fiber optic cable with the following characteristics:

E = 62,000 N/mm^2

d = 127 µm

R_m = 3.18 mm.

These curves show that when the pitch increases, the flexural stress decreases, and the bend radius increases. At a 12.7 cm lay, the flexural stress is about 30 N/mm^2. This value is low compared with the typical minimum tensile strength of fiber.

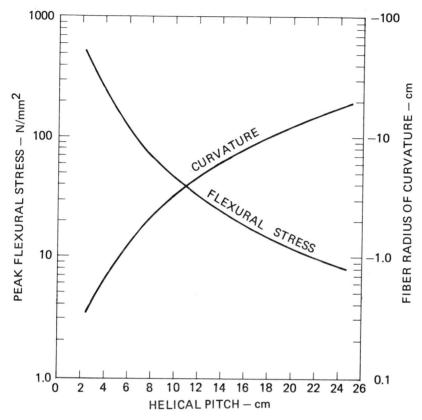

Figure 2.11.   Bend radius and flexural stress versus helical pitch.

2.2.2.3 Strengthening

For applications requiring high tensile strengths (>6N), strength members must be incorporated in the cable.

To maximize a cables tensile stress (strength/unit area), materials with as high a tensile stress as possible at the maximum allowable fiber strain is required. However, the choice of materials for strength members also depends on the required bending radius of the cable, permissible stress levels to which the core may be subjected during cable manufacture, and cost. Furthermore, to achieve high flexibility, multifilament yarn strength

members must be used in contrast to strength members made from
solid materials. Among materials used in optical cables are
Kevlar® Yarn and steel wires. The stress-strain curves of these
materials are shown in Figure 2.12 along with that of a typical
glass fiber for comparison. Where high flexibility is not
required, epoxy glass is often used.

Kevlar® is a high modulus material that can be obtained
as a multifilament yarn. For some applications Kevlar® is coated
with polyurethane. The coating increases the strength and life
of the Kevlar® bundle by reducing filament abrasion and holding
broken filaments in place. Kevlar® is available in two grades —
Kevlar® 49 and Kevlar® 29. As shown in Figure 2.12, Kevlar® 49
exhibits a lower percentage elongation than Kevlar® 29 under identical load conditions, and it is, therefore, better suited for
application as a strength member material in fiber optic cables.
Steel has a higher Young's modulus than Kevlar®. Therefore,
steel provides more strength than Kevlar® for applications where
the allowable fiber strain is below steel's tensile strain, 1.4%.
However, for applications requiring a nonconducting strength member, it cannot be used. In general, epoxy glass has a lower modulus than glass; its modulus depends on the proportion of epoxy
used. The ultimate strain of epoxy glass ranges from 2 to 4%.

2.3   BASIC CABLE CONFIGURATIONS

Optical fiber cables are now produced using a variety
of configurations. These are generally classified as "internal"
or "external" with regard to a strengthening approach; "round" or
"ribbon" with regard to geometry; and as "loose" or "tight" with
regard to construction. Negligible cabling-induced attenuation
has been achieved with all of the designs to be described, at
least in some instances, so a detailed discussion of the attenuation results will not be given. For all designs (to varying

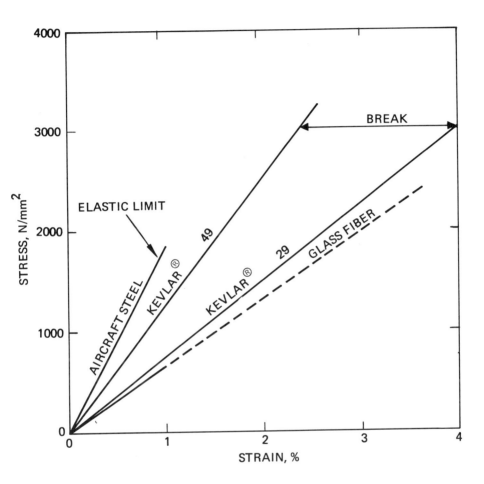

Figure 2.12. Stress versus strain for steel and Kevlar® strength members.

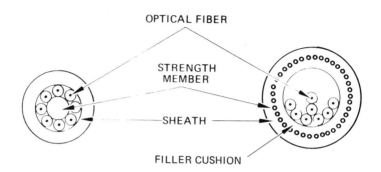

OPTICAL FIBER

STRENGTH
MEMBER

—SHEATH —

FILLER CUSHION

INTERNAL STRENGTH MEMBERS   EXTERNAL STRENGTH MEMBERS

Figure 2.13.   Basic constructions used for optical fiber cables.

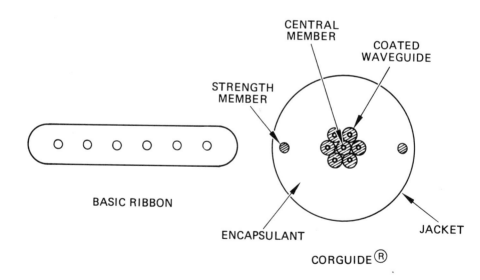

CENTRAL
MEMBER

COATED
WAVEGUIDE

STRENGTH
MEMBER

BASIC RIBBON

ENCAPSULANT   JACKET

CORGUIDE®

Figure 2.14.   Ribbon geometries.

degrees), it is necessary to have smooth internal surfaces, to configure the cable so bending will not cause fiber buckling, and to manufacture the cable to avoid component shrinkage, or at least ensure that it does not lead to fiber buckling. As the technology advances, improvements in cabling techniques should make it possible to select construction, geometry, and strengthening approaches to meet mechanical rather than optical constraints.

The internal and external configurations are illustrated in Figure 2.13. Hybrid configurations are also used. Until recently, external-strength member designs were usually favored. However, new internal-strength member designs are now emerging. Eventually, the approach selection is likely to depend solely on the desired mechanical characteristics of the cable. (Some illustrative examples will be given in the next section.)

Figure 2.14 shows the basic ribbon geometry. Also shown is the Corguide® construction, which is really a hybrid of the round and ribbon geometries. The round geometry offers geometric and bending symmetry; the ribbon design offers ease of fiber identification. The Corguide® geometry permits cable jacketing and strength member application to be performed in a single manufacturing step. In practice, asymmetry of bending has not proved to be a major problem because the cable tends to rotate if it is bent in the strength member plane.

The loose cable construction (2.6) is illustrated in Figure 2.15. Here the fibers are inserted into slots to isolate them from mechanical stress. With this approach, low cabling-induced attenuation is readily achieved. However, the cable is somewhat larger than similar cables with tight construction.

2.4   REPRESENTATIVE CABLES

Numerous performance characteristics must be considered in designing or selecting optical cables.

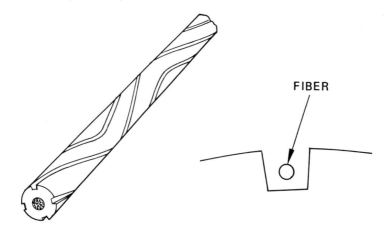

FIBER

Figure 2.15. Loose cable construction (BNR).

        In addition to the uninstalled optical performance
criteria of attenuation, dispersion, source light collection efficiency, and splice/connector losses caused by diameter fluctuations, designers must consider factors related to performance
in the operating environment. Depending on the application, these
performance criteria include

                Tensile strength
                Crush resistance
                Impact resistance
                Flexibility
                Minimum bend radius
                Bending fatigue
                Water/gas blockage
                Torque balance
                Fungus resistance
                Animal bite resistance
                Chemical resistance

For many of these performance criteria, temperature factors must
be addressed also.

A variety of cables that have been designed to meet
specific performance requirements are now available, and additional designs are in various stages of development. In this
section, illustrative examples of current commercial and develop-
mental cables are given, and some of their key characteristics
are described.

Figures 2.16 and 2.17 show single-fiber and two-fiber
industrial cables, respectively. The cables, although relatively
small, are rugged. They can withstand installation loads of
200 Newtons and have survived repeated crossings by large trucks.
Also, they can be installed with relative ease. The two-fiber
cable can be readily split into two single-fiber cables. Thus, it
can be used with single-way connectors and thereby connected to
separated modules.

For outside applications, more rugged cables are
required. Figure 2.18 shows a tight construction cable used for
aerial application, and Figure 2.19 shows a more rugged version
for direct burial. These cables were used in a 23 km mixed
aerial/direct burial/duct installation by Commonwealth Telephone
in western Pennsylvania. The cables were installed and spliced
by telephone company linemen after a brief training period.
Within experimental limits, no degradation of the cables was
observed after installation.

The outer metal tape for the cable shown in Figure 2.19
is provided to increase resistance to damage caused by ground
movement, to prevent water ingress, and to provide some degree
of rodent protection. The voids for the cables of Figures 2.17
and 2.18 are filled with a high viscosity fluid to prevent water
ingress.

The concern with water is a carryover from electrical
cables; the need for such protection has not been proved. With

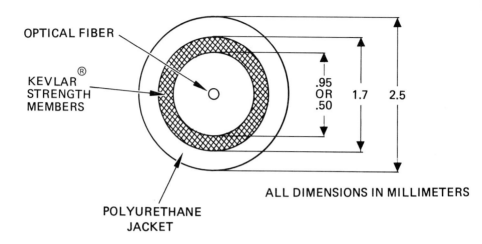

Figure 2.16.   Industrial single fiber cable (ITT).

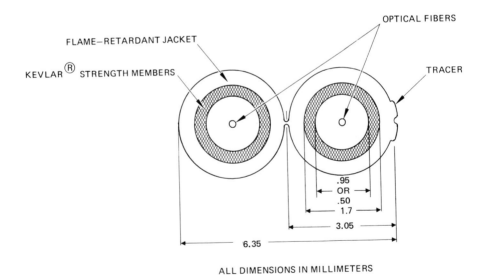

Figure 2.17.   Industrial two fiber cable (ITT).

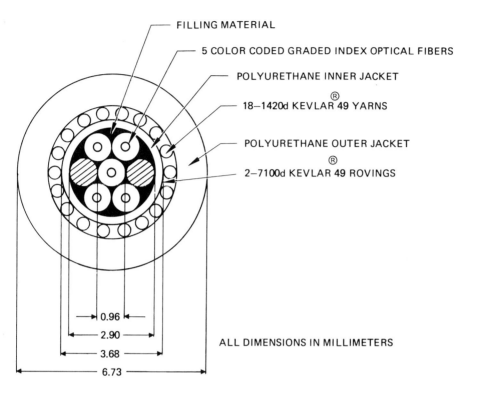

FILLING MATERIAL

5 COLOR CODED GRADED INDEX OPTICAL FIBERS

POLYURETHANE INNER JACKET

18–1420d KEVLAR® 49 YARNS

POLYURETHANE OUTER JACKET

2–7100d KEVLAR® 49 ROVINGS

0.96

2.90

3.68

6.73

ALL DIMENSIONS IN MILLIMETERS

Figure 2.18.   Optical fiber cable for aerial installation(ITT).

regard to subterranean animal protection, the common means is to
use a metal jacket. The aluminum tape only provides protection
against small animals. For larger animals, steel is usually
employed.

The aerial cable has survived a 1,500 N tensile load for
two hours without degradation. It has also survived testing in
accordance with MIL Test C–13777 at –45°C, 25°C, and 85°C as
follows:

Impacts	200 at 4 Nm
Twist	1,000 cycles
Bend	1,000 cycles

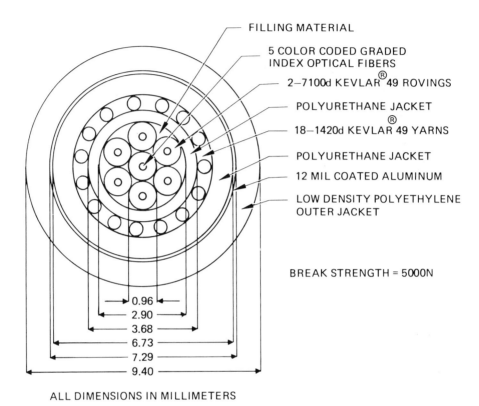

FILLING MATERIAL

5 COLOR CODED GRADED
INDEX OPTICAL FIBERS

2–7100d KEVLAR® 49 ROVINGS

POLYURETHANE JACKET

18–1420d KEVLAR® 49 YARNS

POLYURETHANE JACKET

12 MIL COATED ALUMINUM

LOW DENSITY POLYETHYLENE
OUTER JACKET

BREAK STRENGTH = 5000N

0.96
2.90
3.68
6.73
7.29
9.40

ALL DIMENSIONS IN MILLIMETERS

Figure 2.19. Optical fiber cable for direct burial (ITT).

Both cables exhibit less than 20% change in attenuation
when cycled over a temperature range of –40°C to 60°C and have
passed the Rural Electric Association water blockage test.

Figure 2.20 shows another cable intended for telephone
applications. This cable uses both a fiber glass epoxy central
strength member and Kevlar® external strength members. The
fibers are incorporated in a filled, loose-tube configuration.

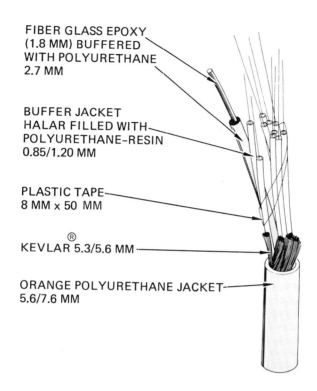

FIBER GLASS EPOXY
(1.8 MM) BUFFERED
WITH POLYURETHANE
2.7 MM

BUFFER JACKET
HALAR FILLED WITH
POLYURETHANE-RESIN
0.85/1.20 MM

PLASTIC TAPE
8 MM x 50 MM

KEVLAR® 5.3/5.6 MM

ORANGE POLYURETHANE JACKET
5.6/7.6 MM

Figure 2.20. Loose tube cable (Siecor).
Performance data are as follows:

Strength	15,000 N
Weight	640 N/km
Operating temperature range	−20 to 50°C
Recommended installation load	≤1,500 N
Bending radius (unloaded)	>75 mm
Crush resistance	≤1,000 N/cm
Impact resistance (13 mm radius hammer)	50 at 3 Nm

The ribbon geometry lends itself to gang splicing. Figure 2.21
shows a cable (2.7) consisting of 12 tapes, each with 12 fibers,
in a crossply jacket. Performance data for this cable are as
follows:

Outside diameter	1.22 cm
Weight	1,170 N/km
Tensile load	2,700 N (assume 344 N/mm^2 diameter fiber)
Bend radius	23 cm
Impact resistance	400 at 6.44 Nm (1.27 cm diameter mandrel)
Cyclic twist	200 cycles (±180°)
Cyclic bend	1,400 cycles ±90° over 12.4 cm radius

The cable shown in Figure 2.22 is designed for ocean bottom links. The cable consists of a single fiber surrounded by helically laid copper-clad steel wires. The wires provide strength and a source of power for repeaters.

The cable exhibits less than 0.1 dB/km attenuation increases at static pressures corresponding to at least 3 km depth in water. Other test results on the cable were as follows:

Length	7.7 km
Attenuation	
0.83 μm	3.7 dB/km
1.06 μm	1.6 dB/km
Bandwidth	
0.90 μm	950 MHz/km
Break strength	490 N
DC resistance	∿38 Ω/km
Test voltage	2,000 Vdc
Flexure life* (cycles)	
10 cm diameter sheave	460,000 minimum
100 N bias load	710,000 average
±28° flexure	1,200,000 maximum

---

*Failure always occurred in the Cu-clad steel wires before the glass fiber broke.

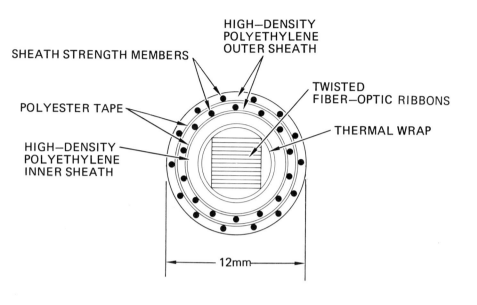

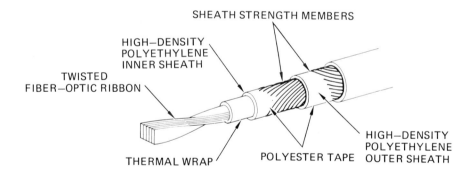

Figure 2.21. Stacked ribbon cable (Western Electric).

Attenuation change over
temperature

-55°C to 80°C   <0.1 dB/km

Experimental cables of underwater vehicle towing have
also been produced. For deep tow applications, cables must sup-
port heavy loads, carry power to the towed vehicle, and carry
communication signals to and from the towed vehicle.

Figure 2.23 shows such a cable. This cable has external
steel strength members to provide protection as well as tensile
strength. The strength member wires are contrawound to improve
torque balance. The design tensile strength of the cable is
147 N. It is designed to carry 10 A and to sustain 3,000 V.

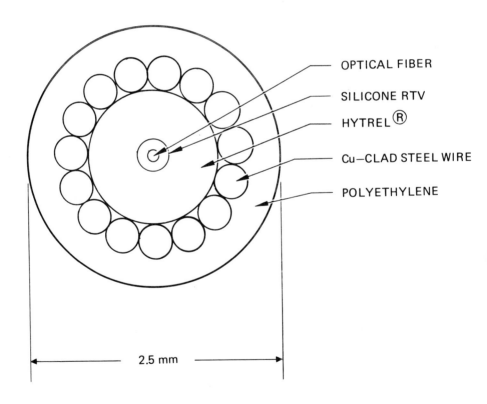

OPTICAL FIBER

SILICONE RTV

HYTREL®

Cu—CLAD STEEL WIRE

POLYETHYLENE

2.5 mm

Figure 2.22. Bottom laid cable.

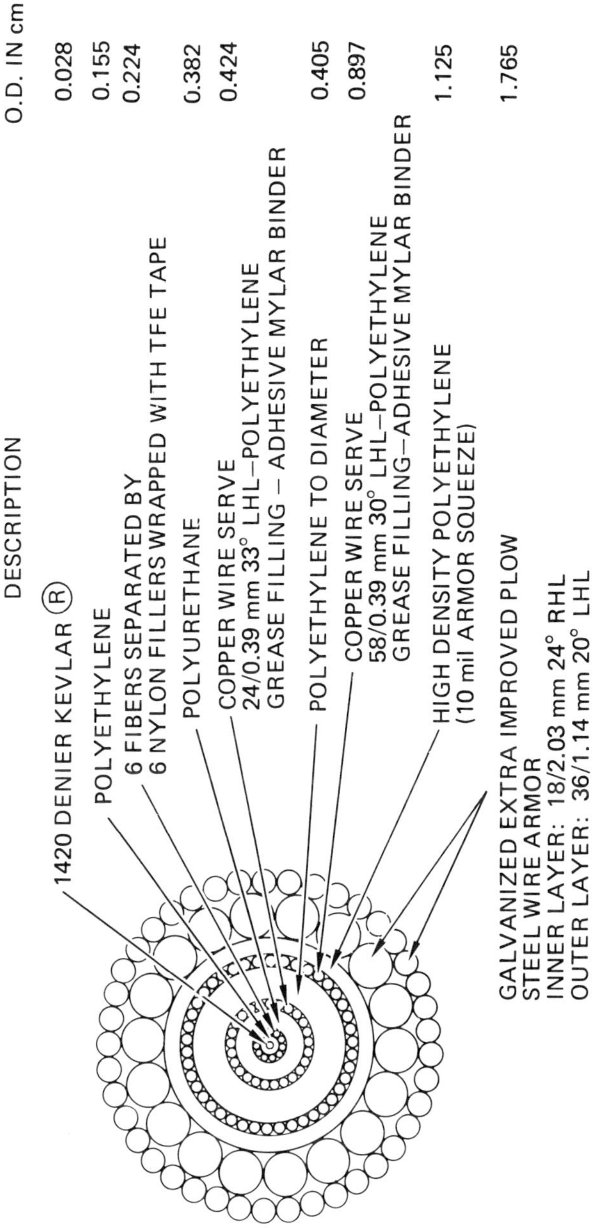

DESCRIPTION	O.D. IN cm
1420 DENIER KEVLAR ®	0.028
POLYETHYLENE	0.155
6 FIBERS SEPARATED BY 6 NYLON FILLERS WRAPPED WITH TFE TAPE	0.224
POLYURETHANE	0.382
COPPER WIRE SERVE 24/0.39 mm 33° LHL–POLYETHYLENE GREASE FILLING – ADHESIVE MYLAR BINDER	0.424
POLYETHYLENE TO DIAMETER	0.405
COPPER WIRE SERVE 58/0.39 mm 30° LHL–POLYETHYLENE GREASE FILLING–ADHESIVE MYLAR BINDER	0.897
HIGH DENSITY POLYETHYLENE (10 mil ARMOR SQUEEZE)	1.125
GALVANIZED EXTRA IMPROVED PLOW STEEL WIRE ARMOR INNER LAYER: 18/2.03 mm 24° RHL OUTER LAYER: 36/1.14 mm 20° LHL	1.765

Figure 2.23. Tow cable.

ACKNOWLEDGMENTS

Much of the work described is freely quoted from previous
publications and direct communications. I wish to acknowledge
M. Pomerantz of CORADCOM; R.L. Eastley of NOSC; and A. Asam,
J.C. Smith, G. Bickel, R. Thompson, P. Oh, C. Schlef, F. Akers,
and C. Kao of ITT, whose contributions to cable technology served
as a basis for this chapter. Also, I thank B. Bielowski of Siecor
for providing data on the loose tube cable. Finally, I gratefully acknowledge the assistance of J. Cole in preparing the
manuscript.

PROBLEMS FOR CHAPTER 2

1.   For a step index fiber, what is the attenuation caused
          by 1,000 bumps/km, assuming Eq. (2.1) applies? Assume
          core radius = 25 μm, fiber diameter = 125 μm, NA = 0.2,
          E_m = 690 N/cm^2, E_s = 62,000 N/mm^2.

2.   Using Figure 2.2 as a basis, what is the cabling-induced
          attenuation for a fiber with a 0.23 mm jacket for each
          jacket type?

3.   What is the Fresnel loss per fiber end for a 0.2 NA
          silica-clad fiber?

4.   By what factor is the time to failure for a given fiber
          crack extended if the fiber stress is reduced two thirds,
          assuming N = 24? N = 15?

5.   Calculate the bending radius and maximum surface stress
          for a helically laid fiber assuming the cable core diam-
          eter = 2.54 mm, the fiber diameter = 125 μm, the pitch
          = 7.6 cm, and E = 62,000 N/mm^2. What is the stress if
          the pitch is reduced to 2.54 cm for this 1 mm diameter
          fiber jacket?

REFERENCES

2.1   R. Olshansky, "Distortion Losses in Cable Optical Fibers,"
      Appl. Opt. 14, No. 1, January 1975, pp. 20-21.

2.2   D. Gloge, "Optical Fiber Packaging and Its Influence on
      Fiber Strengthness and Loss," *Bell Syst. Tech. J. 54,* No. 2
      (1975), p. 245.

2.3   P.R. Bark and U. Oestreich, "Stress-Strain Behavior of
      Optical Fiber Cables," *Optical Fiber Communications* (Topical
      Meeting Digest), Washington, D.C., March 6-8, 1979.

2.4   D. Pinnow, G.D. Roberson, G.R. Blair, and J.A. Wysocki,
      "Advances in High-Strength Metal-Coated Fiber Optical
      Waveguides," *Optical Fiber Communications* (Topical Meeting
      Digest), Washington, D.C., March 6-8, 1979, pp. 28-30.

2.5   M.S. Maklad, A.R. Asam, and F.I. Akers, "Recent Advances
      in High Strength Optical Fibers Having Surface Compression,"
      Proc. 29 Intl. Wire and Cable Symp., November 13-15, 1979,
      pp. 340-343.

2.6   F.O. King and F.P. Kapron, "Rugged Optical Fiber Cables
      Having Attenuation Below 3 dB/km," *Optical Fiber Communi-
      cation* (Topical Meeting Digest), Washington, D.C., March
      6-8, 1979, pp. 28-30.

2.7   P.F. Gagen and M.R. Santana, "Design and Performance of a
      Crossply Lightguide Cable Sheath," *Proc. 28th Intl. Wire
      and Cable Symp.*, Nov. 13-15, 1979, p. 395.

2.8   C. Kao, "Optical Fibers and Cables," Chapter in *Optical
      Fiber Communications*, by M.J. Howest and D.V. Morgan
      (John Wiley and Sons, 1980), pp. 189-249.

CHAPTER 3

COUPLING COMPONENTS FOR OPTICAL FIBER WAVEGUIDES

M. K. Barnoski

TRW   Technology Research Center
El Segundo, California 90245

3.1   INTRODUCTION

The advent of optical fiber waveguides with attenuation
as low as 0.2 dB/km and the rapid advancement in the development
of high brightness LEDs and injection laser diodes (ILDs) have
motivated the investigation of techniques to maximize the coupling
of the radiation emitted by these diodes into multimode fibers.
The efficient coupling of radiation between two fibers to be connected is also being investigated intensely. Presented in this
chapter is an elementary review of coupling radiation emitted by
the source into the fiber waveguide and a review of the fiber-to-
fiber coupling.

3.2   OPTICAL COUPLING INTO GLASS FIBER MULTIMODE
         WAVEGUIDES

3.2.1   Power Transfer Between Emitting and Receiving
         Surfaces

An estimate of the input coupling efficiency of the
source to the fiber can be obtained by considering the power
transfer efficiency between a radiating surface of area A_S and a
receiving surface of area A_R. The power transfer can be computed

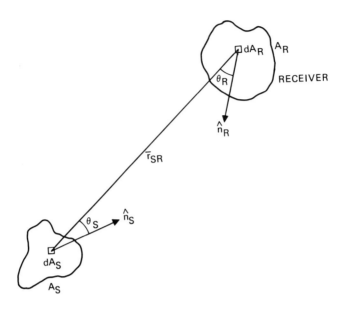

Figure 3.1. Arbitrary orientation of emitting and receiving
             surfaces.

if the goniometric characteristics of the two surfaces are known.
Consider the two arbitrarily oriented surfaces separated by distance vector $\overline{r}_{SR}$, as shown in Figure 3.1. The power transfer from
an element of area dA_S to an element of area $\cos\theta_R\, dA_R$ normal to
r_{SR} is

$$dP_{SR} = B_\Omega(\overline{X}_S, \theta_S)\, dA_S\, d\Omega_R \quad , \qquad (3.1)$$

where $B_\Omega(\overline{X}_S, \theta_S)$ is the brightness of the source in units of watts
per square centimeter steradian. For sources with nonuniform
radiation distributions, B_Ω is a function of $\overline{X}_S$, which is the
position vector on the surface A_S. The quantity $d\Omega_R$ is the solid
angle subtended by $(\cos\theta_R\, dA_R)$ at the source point dA_S, that is,

$$d\Omega_R = \frac{\cos\theta_R \, dA_R}{r^2_{SR}} \, .$$

(3.2)

The total power transfer between A_S and A_R is, therefore,

$$P_{SR} = \int_{A_S} \int_{A_R} B_\Omega(\overline{X}_s, \theta_s) \, \cos\theta_R \, r^{-2}_{SR} \, dA_R \, dA_S \, .$$

(3.3)

In general, this integral is complicated with the angles θ_S and θ_R and the separation vector r_{SR} being dependent on the position coordinates on the surfaces A_R and A_S. To obtain an estimate of the input coupling efficiency, approximations can be made that considerably simplify the integration.

Reference to the two arbitrarily oriented differential surface areas in Figure 3.2 reveals that the area $dA_R \cos\theta_R$, which is normal to the vector $\overline{r}_{SR}$, is approximately equal to the

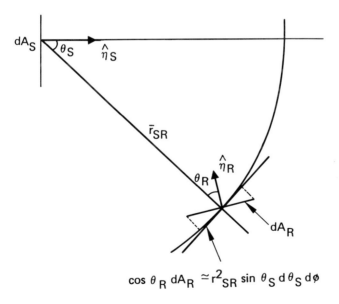

$$\cos\theta_R \, dA_R \simeq r^2_{SR} \sin\theta_S \, d\theta_S \, d\phi$$

Figure 3.2. Arbitrarily oriented differential surface areas, dA_S and dA_R.

differential area of a sphere centered at dA_S, i.e.,
$dA_R \cos\theta_R \simeq r_{SR}^2 \sin\theta_S \, d\theta_S \, d\phi$. The angle ϕ is an angle varying
from 0 to 2π in the source plane. The total power transferred
between A_S and A_R is, therefore,

$$P_{SR} = \int_{A_S} \int_{\theta_S} \int_{\phi} B_\Omega(\overline{X}_S, \theta_S) \sin\theta_S \, d\theta_S \, d\phi \, dA_S \quad . \tag{3.4}$$

The integration is considerably simplified if the source
has a uniform radiation distribution across its area and if the
extent of the source area is small compared with the separation
vector r_{sR}. When these conditions are satisfied, the total power
transfer becomes

$$P_{SR} = 2\pi A_S \int B(\theta) \sin\theta \, d\theta \quad , \tag{3.5}$$

where cylindrical symmetry has also been assumed.

3.2.2   Goniometric Characteristics of Opto-Electronic Sources

Two sources of interest for use in fiber optic communications systems are the semiconductor light-emitting diode (LED)
and the semiconductor injection laser diode (ILD). State-of-the-
art, high brightness LEDs and cw room-temperature, injection
lasers are commercially available from numerous suppliers. The
emission areas of these devices are well matched to the core area
of a single fiber. High-brightness LEDs can be flat geometry
surface emitters, such as those developed by Burrus (3.1) or strip
geometry edge emitters developed at RCA laboratories (3.2). (A
cross-sectional drawing of a double heterojunction electroluminescent diode coupled to a single optical fiber is shown in
Figure 4.17 of Chapter 4.) The 50-μm-diameter emission area of
this device is well matched to the core area of the low-loss
fiber.

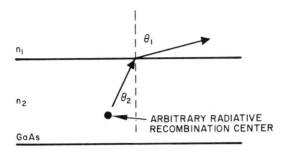

Figure 3.3. Planar-type GaAs LED schematic.

The angular distribution of radiation emitted from the surface of such an LED can be determined with the aid of the schematic diagram shown in Figure 3.3.

The radiation from any arbitrary point interior to the LED is isotropic into the full 4π steradians; however, because of refraction at the interface of the LED (whose refractive index is n_2) and the surrounding medium of index n_1, the radiation emerging from a planar geometry LED has an angular dependence.

Let the photometric intensity be I(watts/steradian). The conservation of energy requires that

$$I_2 \, d\Omega_2 = I_1(\theta_1) \, d\Omega_1 \qquad (3.6)$$

or

$$I_2 \, \sin\theta_2 \, d\theta_2 \, d\phi = I_1(\theta_1) \, \sin\theta_1 \, d\theta_1 \, d\phi \, , \qquad (3.7)$$

which, using Snell's law of refraction,

$$n_2 \, \sin\theta_2 = n_1 \, \sin\theta_1 \, , \qquad (3.8)$$

and its differential

$$n_2 \cos\theta_2 \, d\theta_2 = n_1 \cos\theta_1 \, d\theta_1 \quad , \tag{3.9}$$

yield

$$I_1(\theta_1) = \left(\frac{n_1}{n_2}\right)^2 I_2 \frac{\cos\theta_1}{\cos\theta_2} \quad . \tag{3.10}$$

Expressing θ_2 in terms of θ_1 and substituting into Eq. (3.10) results in

$$I_1(\theta_1) = \left(\frac{n_1}{n_2}\right)^2 I_2 \frac{\cos\theta_1}{\left[\left(1-\left(\frac{n_1}{n_2}\right)^2 \sin\theta_1^2\right)\right]^{1/2}} \quad , \tag{3.11}$$

which since

$$\frac{n_1}{n_2} < 1 \quad , \tag{3.12}$$

becomes approximately

$$I(\theta) \approx \left[\left(\frac{n_1}{n_2}\right)^2 I_2\right] \cos\theta \quad . \tag{3.13}$$

The external radiant intensity from a planar geometry surface-emitting LED is the same as a planar radiating surface with brightness of value

$$B(\theta) = B \cos\theta \tag{3.14}$$

with

$$B = \left[\left(\frac{n_1}{n_2}\right)^2 \frac{I_2}{A_{source}}\right] \quad , \tag{3.15}$$

where A_{source} is the emitting area of the LED and I_2 depends on drive current, quantum efficiency, and reflectivity. For ranges

of θ of interest, the internal reflectivity is approximately
constant; consequently, I_2 can be considered constant. The radiation pattern obtained from a 50 μm active diameter, surface-
emitting LED manufactured by Plessey Opto Electronics is shown
in Figure 3.4. As can be seen, the distribution approximates a
cosine distribution, as was predicted for a planar radiating surface. A surface that radiates with a cosine distribution is a
Lambertian radiator.

A more directional beam can be obtained from an edge-
emitting, double heterojunction LED. This beam results when the
active p-n junction region of the device is sandwiched between
two layers of semiconductor material with a refractive index less
than that of the active region. The resulting optical waveguiding
yields a more directional, external radiation pattern. (Typical

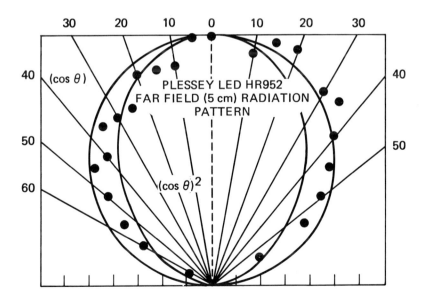

Figure 3.4. Angular radiation pattern of Plessey HR952 LED.

radiation patterns obtained from an edge-emitting, electroluminescent diode are shown in Figure 4.18 of Chapter 4. Also shown
in the figure are the radiation patterns obtained from a double
heterojunction laser diode. As can be seen, the emission pattern
obtained from the laser diode is more well directed than that of
the LED.)

3.2.3   Acceptance Angle of the Fiber

The amount of energy coupled into a fiber is greatly
dependent on its numerical aperture (NA), which, for a step index
fiber, can be defined with the aid of Figure 3.5. For total
internal reflection to occur, the internal reflection angle must
be greater than the critical angle $\sin\phi_c = n_1/n_2$. This can be
related to the incident angle θ via Snell's law

$$n_o \sin\theta = n_2 \sin\psi = n_2 \sin(90 - \phi_c) = (n_2^2 - n_1^2)^{1/2} \quad .$$

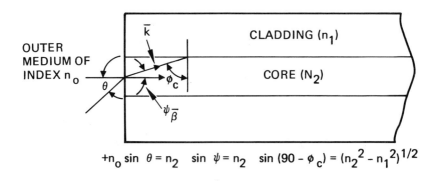

$$+n_o \sin \theta = n_2 \sin \psi = n_2 \sin (90 - \phi_c) = (n_2^2 - n_1^2)^{1/2}$$

Figure 3.5.   Entrance angle of cladded fiber for total internal
reflection at core-cladding interface.

The quantity n_o $\sin\theta$ is the NA of the fiber. When the outer medium is air, NA = $\sin\theta \simeq \theta$ for small NA. Since the fiber accepts only those rays contained within a cone whose maximum angle is determined by total internal reflection at the core-clad interface, an input coupling loss will result if the angular emission cone of the source exceeds that defined by the NA of the fiber. This is illustrated schematically in Figure 3.6 for a source coupled to a fiber. Only those rays contained within the shaded cone are trapped in the core.

The optical ray illustrated in Figure 3.5 is a meridional ray, i.e., a ray which passes through the axis of the waveguide at some part of its path. In addition to meridional rays, there are also skew rays that do not intersect the fiber axis. A plane wave incident on the core of the fiber excites both meridional and skew rays. Meridional rays are not excited if the irradiance angle θ exceeds the NA of the fiber. Skew rays, however, can be excited

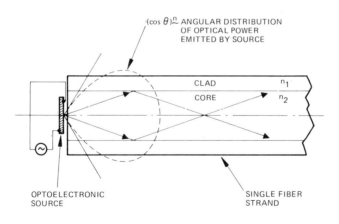

$$P_{FIBER} \simeq P_{SOURCE} \left(\frac{n+1}{2} \right) N.A.^2$$

$$N.A.^2 = (n_2{}^2 - n_1{}^2)^{1/2}$$

Figure 3.6.   Schematic diagram of source coupled to single-strand fiber waveguide.

beyond the numerical aperture of the step index fiber (3.3).
These high-angle rays are often neglected because they are usually highly attenuated. The power carried by skew rays will be
neglected in the following analysis of coupling efficiency.

3.2.4   Coupling Efficiency

3.2.4.1 Direct Coupling

If the opto-electronic source is directly butted against
the fiber, the input coupling factor can be determined directly.
The power coupled into a step index fiber is

$$P_{fiber} = 2\pi \, A_{source} \int_{o}^{NA} B(\theta) \, \sin\theta \, d\theta \quad , \qquad (3.16)$$

where A_{source} is the emitting source area. For the case where
the angular distribution of the source brightness can be expressed
as $B(\theta) = B(\cos\theta)^{n}$ for $|\theta| \leq \theta_{S}$ and zero for all other θ, the
above integral is easily evaluated. The optical power coupled
into the fiber is

$$P_{fiber} = 2\pi \, A_{source} \, B \int_{o}^{NA} (\cos\theta)^{n} \, \sin\theta \, d\theta \qquad (3.17)$$

or

$$P_{fiber} = 2\pi A_{source} \, B \left[\frac{1-(\cos \, NA)^{n+1}}{n+1} \right] \quad . \qquad (3.18)$$

The total optical output power emitted from the source into the
full hemisphere is given by

$$P_{source} = 2\pi \, A_{source} \, B \int_{o}^{\theta_{S}} (\cos\theta)^{n} \, \sin\theta \, d\theta \quad , \qquad (3.19)$$

or

$$P_{source} = 2\pi \ A_{source} \ \frac{B}{n + 1} \ \left(1 - (\cos\theta_s)^{n+1}\right) \ . \qquad (3.20)$$

The power coupled into the fiber can, therefore, be expressed in terms of the total optical power emitted from the source, i.e.,

$$P_{source} \left(\frac{1 - (\cos \ NA)^{n+1}}{1 - (\cos\theta_s)^{n+1}}\right) \ , \qquad (3.21)$$

which for small NA and $\theta_s = \pi/2$ reduces to

$$P_{fiber} = P_{source} \left(\frac{n + 1}{2}\right) \ NA^2 \ . \qquad (3.22)$$

It should be noted that the assumption has been made that the cross-sectional area of the radiation pattern at the entrance plane of the fiber is less than or equal to the fiber core area (i.e., no area mismatch loss).

The direct butt coupling efficiency for light coupled into a graded index fiber can be treated similarly. The index of refraction of the core of a graded index fiber varies with position on the face of the core. The refractive index variation with radius will be taken to be the power law

$$n(r) = n_2^2 \left[1 - 2\Delta \left(\frac{r}{a}\right)^\alpha\right] \ , \qquad (3.23)$$

where $n_2 = n(o)$ is the refractive index at the center of the fiber core, α specifies the shape of the refractive index profile, and $\Delta = (n_2^2 - n_1^2)/2n_2^2$ with $n_1 = n(a)$ the refractive index at the edge of the fiber core. Since the index varies with radius, the fiber acceptance angle is also dependent on the radial variable r. The acceptance angle can be determined by phase matching the incoming ray at radial position r to the ray (wavevector $\bar{\beta}$) associated with the bound mode at cutoff at fiber radius r. The phase-match condition between the refracted incoming ray with wavevector

$\bar{k}$ and bound mode with wavevector $\bar{\beta}$ (refer to Figure 3.5) is given
by

$$|\bar{\beta}| = |\bar{k}| \cos\psi \quad . \tag{3.24}$$

At radial position r the angle of incidence is related to the
angle of refraction by Snell's law,

$$n_o \sin\theta = n(r) \sin\psi \quad . \tag{3.25}$$

Using Eqs. (3.24) and (3.25) and $\beta = n_g k_o$ and
$k = n(r) k_o$, the radially dependent critical acceptance angle, by
setting β equal to its cutoff value $\beta = n_1 k_o$, can be expressed as

$$\theta_c(r) = n(r) \arccos\left(\frac{n_1}{n(r)}\right) \quad . \tag{3.26}$$

Here n_g is the effective index of the guide mode and $k_o = 2\pi/\lambda_o$
is the free-space wavevector. As for the step index case, it is
assumed that the outer medium is air ($n_o = 1$). By using the
integral expression (Eq. (3.4)), the optical power coupled into
the graded index fiber is

$$P_{fiber} = \frac{4\pi^2 B}{n+1} \int_o^a r\left(1 - \cos_{\theta_c}^{n+1}(r)\right) dr \quad , \tag{3.27}$$

which upon substitution and integration becomes (small NA and
$\theta_s = \pi/2$)

$$P_{fiber} = \frac{1}{2}(n+1) NA^2 \frac{\alpha}{\alpha+2} P_s \quad . \tag{3.28}$$

For a step index fiber α is infinity. In this case, Eq. (3.28)
becomes identical to Eq. (3.22), as expected.

The amount of power coupled into both a graded and step
index is proportional to the square of the NA and increases as
the directionality of the source emission pattern improves, i.e.,
as n becomes larger. It is important to note that the amount of

optical power coupled from a given source into a graded fiber with
on axis numerical aperture (NA) is less by a factor $\alpha/\alpha + 2$
than that coupled into a step index fiber with the same NA. For
a fiber with a square law profile $\alpha = 2$. In this case there is a
3 dB reduction in coupling efficiency.

This result can also be obtained by recognizing that the
power coupled into the fiber is directly proportional to the number of bound modes supported by the waveguide. For a step index
guide, the number of bound modes is approximately

$$N = \frac{v^2}{2} \quad , \tag{3.29}$$

where

$$v^2 = (ka)^2 \, (n_2^2 - n_1^2) = (ka)^2 \, NA^2 \quad . \tag{3.30}$$

For a graded fiber the number of bound modes is

$$M = \frac{\alpha}{\alpha+2} \, (ka)^2 \, (n_2^2 - n_1^2) = \frac{\alpha}{\alpha+2} \, N \quad . \tag{3.31}$$

For a given n_2 and n_1, the number of modes in a graded fiber is
smaller by a factor $\alpha/\alpha + 2$ than the number in a step fiber. As
a result, the coupling efficiency is reduced by this factor.

Once again, this analysis includes only the bound or
trapped rays which travel in the direction of the waveguide axis
and which undergo total internal reflection at the core-clad
interface. It neglects the leaky modes that travel obliquely
(skew) to the waveguide axis and undergo only partial reflection
at the core boundary. The leaky modes are suitable approximations for the radiation field within the fiber. In some cases,
a significant portion of the radiation field can persist for long
distances in the fiber waveguide. A complete analysis must,

therefore, include the leaky rays. C. Pask and A.W. Snyder (3.4)
treat this problem in detail using a modified form of geometric
optics.

3.2.4.2 Coupling Using a Lens

The above results obtained with the source placed
directly against the fiber core revealed that the input coupling
coefficient was proportional to the square of the NA of the fiber.
Since the NA is less than unity (typical values range from 0.18
to 0.5), a considerable amount of power emitted by the source can
be lost in input coupling. It is of interest to investigate the
effect on the coupling loss of introducing an intervening optical
element between the radiating source and fiber, as illustrated in
Figure 3.7. For simplicity, it is assumed that the radiation
emanating from the source is a constant (n = 0) within the solid
angle Ω_s; the angular distribution of the source brightness is
assumed to be of the form

$$B(\theta) = \begin{cases} B & \theta \leq \theta \\ 0 & \theta > \theta \end{cases} \quad . \tag{3.32}$$

Again for simplicity, cylindrical symmetry is assumed. A detailed
treatment of coupling both disc and strip geometry devices with
and without a lens where these simplifying approximations are not
made can be found in Ref. (3.5).

The total power in the input cone Ω_1 (the total power
collected by the lens), is

$$P_{lens} = 2\pi \, B \, A_{source} \int_{o}^{\theta_L} \sin\theta \, d\theta \tag{3.33}$$

$$P_{lens} = 2\pi \, B \, A_{source} \, (1 - \cos\theta_L) \quad , \tag{3.34}$$

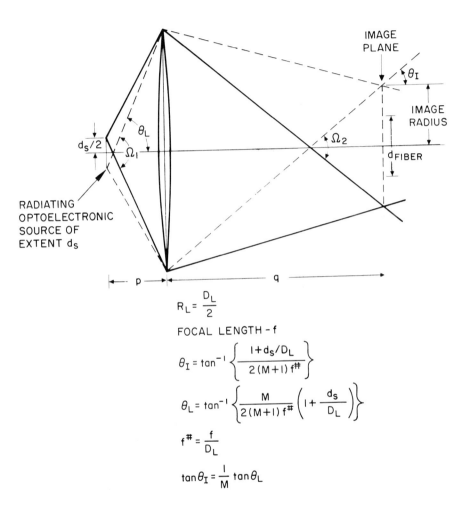

$$R_L = \frac{D_L}{2}$$

FOCAL LENGTH - f

$$\theta_I = \tan^{-1}\left\{\frac{1+d_s/D_L}{2(M+1)f^{\#}}\right\}$$

$$\theta_L = \tan^{-1}\left\{\frac{M}{2(M+1)f^{\#}}\left(1+\frac{d_s}{D_L}\right)\right\}$$

$$f^{\#} = \frac{f}{D_L}$$

$$\tan\theta_I = \frac{1}{M}\tan\theta_L$$

Figure 3.7. Geometry used for lens input coupler calculations.

where

$$\theta_L = \tan^{-1} \left[\frac{M}{2(M+1) \, f^{\#}} \left(1 + \frac{d_s}{D_L} \right) \right] \quad . \tag{3.35}$$

For simplicity, it has been assumed that the thin-lens formula applies (i.e., $1/p + 1/q = 1/f$). The definitions for the quantities, $M, f^{\#}, D_L$ are

> M = magnification = q/p
>
> D_L = lens diameter
>
> f = focal length of lens
>
> $f^{\#}$ = f-number of lens = f/D_L .

The formulas developed herein involving the trigonometric expressions, including those shown in Figure 3.7, are exact to the extent that the thin-lens formula applies.

The angular distribution in the input cone Ω_1 is uniform, provided that the lens collection angle θ_L is less than or equal to the maximum emission angle θ_s from the source, i.e., $\theta_L \leq \theta_s$. For this case, the power collected by the lens must also be uniformly distributed in the output cone Ω_2 where

$$\Omega_2 = 2\pi \int_o^{\theta_I} \sin\theta \; d\theta = 2\pi(1 - \cos\theta_I) \tag{3.36}$$

with

$$\theta_I = \tan^{-1} \left(\frac{1 + \dfrac{d_s}{D_L}}{2(M+1) f^{\#}} \right) \quad . \tag{3.37}$$

The watts/steradian contained in the output cone Ω_2 is, therefore,

$$\frac{P_{lens}}{\Omega_2} = B \; A_{source} \frac{(1 - \cos\theta_L)}{(1 - \cos\theta_I)} \quad . \tag{3.38}$$

The area of the source image (in the image plane of the lens) is $A_I = M^2 A_{source}$. If this image area is less than that of the fiber core,

$$M \leq \frac{d_{fiber}}{d_{source}} \quad , \quad (3.39)$$

then all the rays in the cone Ω_2 with angles less than the NA will be collected by the fiber. The total power in the fiber is, therefore,

$$P_{fiber} = 2\pi \frac{P_{lens}}{\Omega_2} \int_o^{NA} \sin\theta \, d\theta \quad (3.40)$$

$$= 2\pi \, B \, A_{source} \frac{1 - \cos\theta_L}{1 - \cos\theta_I} (1 - \cos_{NA}) \quad .$$

Now, the total power radiated from the source into Ω_s steradians is given by

$$P_{source} = 2\pi \, B \, A_{source} \int_o^{\theta_s} \sin\theta \, d\theta \quad (3.41)$$

$$P_{source} = 2\pi \, B \, A_{source} (1 - \cos\theta_s) \quad , \quad (3.42)$$

where Ω_s is the solid angle that contains the useful external emission being radiated from the emitting diode.

Thus the total optical power coupled into the fiber core with an intervening lens is

$$P_{fiber} = P_{source} \frac{(1-\cos\theta_L)}{(1-\cos\theta_s)} \frac{(1-\cos NA)}{(1-\cos\theta_I)} \quad . \quad (3.43)$$

These calculations show that the input coupling loss can be
reduced to Fresnel losses if the source geometry, fiber core size,
and lens system are selected so that

$$\theta_s \leq \theta_L$$

$$\theta_I \leq NA \qquad\qquad (3.44)$$

$$M \leq d_{fiber}/d_{source}$$

are all satisfied simultaneously. If it is assumed that the
source configuration and lens diameter have been selected so that
$\theta_s = \theta_L$, then

$$P_{fiber} = P_{source} \frac{(1-\cos NA)}{(1-\cos\theta_I)} \quad , \qquad (3.45)$$

which for small angles is approximately given by

$$P_{fiber} \simeq P_{source} \left(\frac{1-\cos NA}{1-\cos \frac{1}{M}\theta_s} \right) \quad , \qquad (3.46)$$

since, in the small angle approximation, $\theta_I \simeq 1/M\ \theta_s$. Clearly,
optimized input coupling occurs when $\theta_s = M(NA)$. The input
coupling loss can be reduced to Fresnel reflection loss if the
angular emission from the source (θ_s) can be sufficiently colli-
mated to have no angle (θ_I) greater than the numerical aperture of
the fiber, while simultaneously maintaining the cross-sectional
area of the radiation pattern at the input to the fiber less than
the core area itself. The results are an illustration of the law
of brightness (3.6), which states that the brightness of the image
cannot exceed that of the object, and can only be equal to it if
the losses of light within the optical system are negligible.

For typical values of θ_s/M, Eq. (3.46) can be written as approximately

$$P_{fiber} = P_{source} M^2 \left(\frac{1-\cos NA}{1-\cos\theta_s}\right) . \qquad (3.47)$$

(For $M = 5$ and $\theta_s = \pi/2$, $1/M^2$ and $(1-\cos(\theta_s/M))$ differ by 18%.) Comparing Eq. (3.48) with that obtained for direct-butt coupling (Eq. (3.21)) reveals that by appropriate selection of the source, fiber, and lens, the input coupling efficiency can be improved over the direct-butt case by the factor

$$M^2 = \left(\frac{d_{fiber}}{d_{source}}\right)^2 = \frac{A_{fiber}}{A_{source}} .$$

Once again, this results because the intervening optical element collimates the angular distribution of source radiation by simultaneously magnifying the radiation source area. Clearly, the coupling efficiency can be improved only if the product of the optical emission area times the emission solid angle is less than the product of the fiber-core area times its solid angle of acceptance. The improvement factor is approximately equal to the magnification squared. In the calculations presented here, isotropic radiation was assumed. The same result can also be obtained (3.7) for a Lambertian ($n = 1$) source.

The coupling efficiency improves as the ratio of the fiber-core area to source-emission area increases. For a given source area more power can be coupled into the fiber as its core area is increased. High input coupling thus dictates a large core. As the core area is increased, the number of bound modes also increases causing a reduction in fiber bandwidth. This problem, or problems encountered with an overly mechanically rigid fiber, limit the maximum core area. Commercially available "fat fibers" have core diameters ranging from 100 to 200 µm. For a

given fiber core size, input coupling efficiency can be improved
by reducing the emission area of the source. The total optical
power emitted by the source is directly proportional (Eq. (3.20))
to the product of the emission area and the brightness. If the
emission area is decreased to enhance coupling efficiency, then
the source brightness must be increased if the total power emitted
from the source is to remain constant. Sources for fiber systems,
therefore, should be as bright as is consistent with restraints
placed on device size by semiconductor materials, device processing, and device lifetime considerations.

The taper launcher (3.8), (3.9) is a simple optical element that has efficiently coupled light from both LEDs and ILDs
into an optical fiber. The element shown in Figure 3.8 consists
of a short section of fiber waveguide heated and pulled into a
taper. The magnification of the taper is equal to the square of
the taper ratio $R = (a/a_1)$. Experimental results have been
obtained with a 22 μm diameter Lambertian LED. Plotted in Figure
3.9 is the coupling efficiency improvement factor (magnification
squared) versus the taper ratio with measured data points also
shown. The experimental results obtained are in excellent agreement with those predicted by theory.

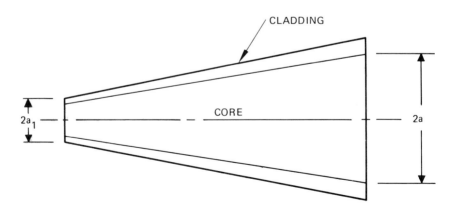

Figure 3.8.   Tapered launcher input coupler.

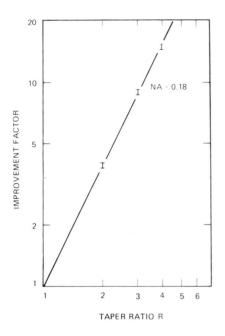

Figure 3.9.
Dependence of the power
coupling efficiency on the
taper ratio for a Lambertian
source (3.9).

Heterojunction LEDs and double-heterojunction lasers are
not Lambertian radiators, but have more directional radiation
patterns. As a consequence, higher coupling efficiency is
expected for these devices for both direct-butt coupling and
coupling with an optical element. For an edge-emitting, double-
heterojunction LED (3.9), direct-butt coupling to 0.18 and 0.26
NA fibers of 11 and 17%, respectively, have been reported. These
values exceed the values of 3.2 and 6.8% obtainable from a
Lambertian radiator. Using a tapered launcher, maximum experimental coupling efficiencies of 53 and 83% have been measured
for 0.18 and 0.26 NA fibers, respectively.

For a double-heterojunction laser diode, direct-butt
coupling efficiency of 30% into a 0.18 NA fiber has been
reported (3.8). The coupling was improved by a factor of 3.2
to 97% using a 4.3 mm long taper with a ratio of 3.4.

Small ball lenses have also been used (3.10) to improve
input coupling efficiency. A self-aligning microlens system used
with a surface-emitting heterostructure LED is shown in Figure
3.10. The spherical lens is placed in the well etched into the
surface of the LED. Since the LED is a surface emitter, its radiation pattern is Lambertian, and, consequently, the maximum
direct-butt coupling efficiency is NA^2 (using Eq. (3.22)). The
maximum attainable coupling efficiency employing the ball lens is
$(d_{fiber}/d_{led})^2$ $(NA)^2$. (The experiments described in (3.10) used
an 80-μm-diameter fiber with a 0.14 NA.)

3.2.5   Sensitivity of Input Coupling to Mechanical Alignments

In the design and development of practical input (source-
to-fiber) coupling connectors for single-strand systems, it is not
only important to realize the expected magnitude of the coupling
coefficient, but also the effects of mechanical alignment tolerances on the input coupling loss. That severe mechanical alignment tolerances are imposed can be seen from the plot shown in
Figure 3.11 of the increase input coupling loss as a function of
radial displacement of the center of a Corning low-loss step
index (NA = 0.14) fiber and the center of the 50-μm-diameter
surface-emitting LED manufactured by Plessey. As can be seen
from the figure, the input coupling is extremely sensitive to
lateral misorientation. For example, to maintain less than a
1 dB increase requires transverse alignment tolerances of less
than ±20 μm. Note also that a ±50 μm misalignment of the center
of the fiber with respect to the center of the LED results in an
8 dB increase in input coupling loss. The effects on input coupling loss of longitudinal fiber-to-source separation and angular
misalignment of the axis of the source to that of the fiber are
shown in Figures 3.12 and 3.13. As can be seen, the input

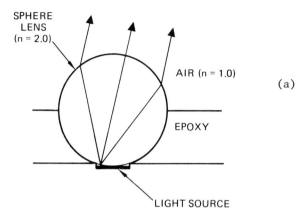

(a)

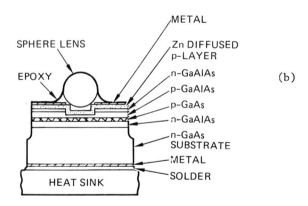

(b)

Figure 3.10. (a) A simple geometrical configuration of the LED
with a sphere lens (3.10). (b) A schematic cross-
sectional drawing of the GaAs–GaAlAs heterostructure
LED with a self-aligned sphere lens.

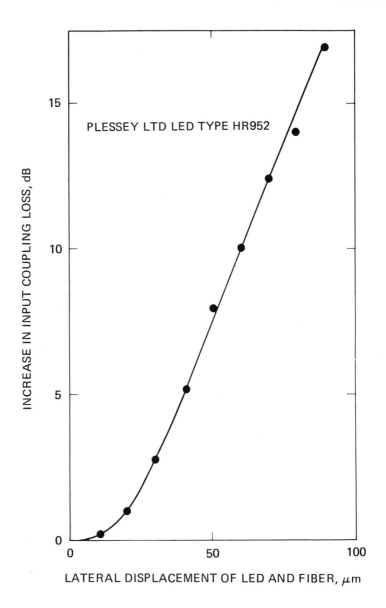

Figure 3.11.   Increase in input coupling loss as a function of
lateral source-fiber displacement.

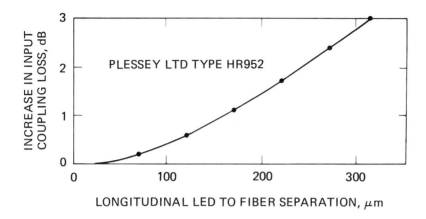

Figure 3.12. Increase in input coupling loss as a function of
longitudinal LED to fiber separation.

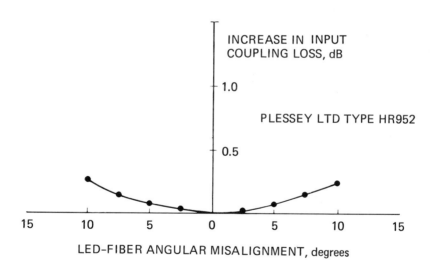

Figure 3.13. Increase in input coupling loss as a function of
angular misalignment.

coupling is relatively insensitive to LED-fiber separation. For
example, with the LED and fiber separated by 150 μm, the increase
in coupling loss is less than 1.0 dB, while an angular tilt of
the fiber axis with respect to the LED surface of 10° increases
the loss by approximately 0.25 dB. This is consistent with the
Lambertian characteristics of the radiation pattern. The data
shown in Figures 3.11, 3.12, and 3.13 indicate, as expected, that
the coupling loss for the surface-emitting LED is most sensitive
to radial misalignment of the LED and fiber core.

3.3   CONNECTORS AND SPLICES

In addition to considering the interconnection of the
electroluminescent sources and receiving photodiodes with the
fiber waveguides, attention must also be directed to the problem
of connecting sections of the optical fiber cable. Practical
system considerations require both low insertion loss connect-
disconnect connectors and permanent splice coupling. The quick
connect-disconnect connector finds application in the interconnection of fiber cables, connection of cables to transmit and
receive modules, and the connection of distribution system components such as serial tee and star couplers that are used in
multiterminal data distribution systems. On the other hand,
environmentally sound, permanent cable splices are required for
long distance, low-loss, data transmission lines. The engineering design of these splice coupling components is dependent on
the packaging (cable type) and number of fibers to be interconnected at the splice.

The attenuation in fiber connectors and splices resulting from intrinsic fiber parameters is dependent, of course, on
the tolerance control that is maintained on these parameters by
the fiber manufacturer and, to some degree, on the actual interconnection technique employed. The attenuation resulting from
parameters extrinsic to the fiber, such as mechanical alignment,

is very dependent on the interconnection technique used. Knowledge of the sensitivity of the connection loss to various types
of fiber end misalignments is required to provide guidance in the
design of connectors and splices. Experimental data measuring
the effect of misalignments on interconnection loss have been
obtained (3.11) using experimental setups such as that shown in
Figure 3.14. The LEDs used as optical power sources were surface
emitters with a 50-μm-diameter emitting area. In the experiment
shown in Figure 3.14(a), a 1.8 m section of fiber was used. The
fiber core diameter was 50 μm with a 100 μm overall diameter.
The experiment shown in Figure 3.14(b) used 20 m of fiber with
55-μm-diameter core and a 100 μm outside diameter. Both fibers
were graded index with on-axis NA of 0.20.

Measured coupling losses in decibels as a function of
normalized mechanical misalignments are shown in Figure 3.15 for
the two experimental cases. The lateral offset d and the longitudinal separation S have been normalized to the fiber core
radius a. The angular misalignment β is normalized to the on-
axis NA. The first experiment refers to the arrangement illustrated in Figure 3.14(a); the second experiment refers to that
of Figure 3.14(b). The data shown in the figure were acquired
by first maximizing the fiber throughput then by cutting the
fiber at the center of its length and aligning the resultant ends
with the micropositioners. In both cases, index matching fluid
was applied to the joints. The power output was measured to be
0.01 dB less than the optimized throughput value measured for
the continuant length in the first experiment and 0.007 dB less
in the second experiment. The differences observed between the
two experiments at zero offset and separation is the result of
these measured differences. The measured results clearly show
that coupling loss is most sensitive to lateral fiber offset.

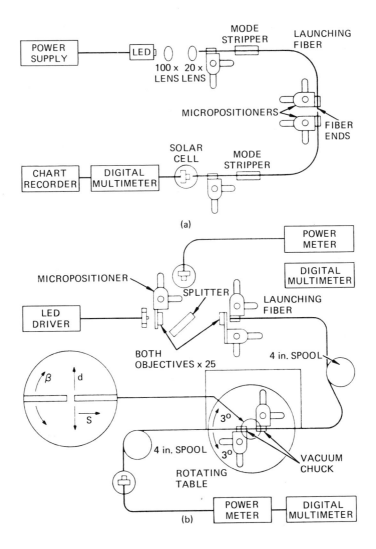

Figure 3.14.   (a) Measurement setup for the first experiment
              (3.11).   (b) Measurement setup for the second
              experiment.

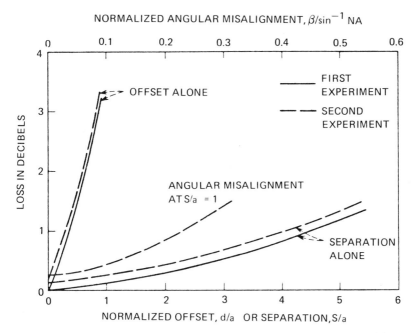

Figure 3.15.   Loss in decibels versus normalized offset d/a, sep-
              aration S/a, and angular misalignment $\beta/\sin^{-1}$ NA
              (3.11).

For example, to maintain a loss of 0.1 dB, a lateral offset of
approximately 1 μm is required for a 55 μm core diameter fiber.
Using the independent measurements made of the coupling loss
dependence on the three normalized parameters, a set of constant
loss curves caused by the various kinds of misalignments can be
constructed.   Figure 3.16 displays constant loss lines resulting
from fiber end offset d/a, separation S/a, and angular misalignment $\beta/\sin^{-1}$ NA.   A coupling loss of 0.5 dB can result from either
a normalized offset of 0.2 alone, a normalized separation of 2.0
alone, or a combination of normalized offset of 0.1 and normalized

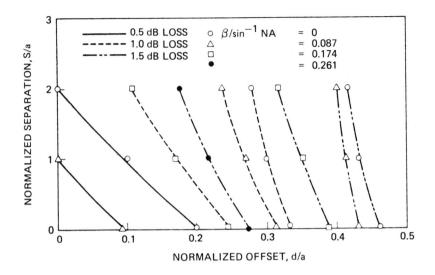

Figure 3.16.   Constant loss lines as the result of fiber and offset d/a, separation S/a, and angular misalignment
$\beta/\sin^{-1}$ NA (3.11).

separation of 1.   The experimentally generated curves shown for a
graded index fiber provide a valuable guide to the connector and
splice designer.   The losses shown in the figure are the minimum
losses since they only include effects of mechanical misalignment
and not fiber intrinsic losses or effects of improperly prepared
or cleaned fiber ends and Fresnel reflection loss.   Fresnel
reflection losses are approximately 0.2 dB per fiber face.   The
Fresnel loss can, therefore, be as large as 0.4 dB for a "dry"
connector with an air gap.   This loss, of course, can be minimized by using index matching materials.   Elimination of the
effect of end preparation and cleanliness is related to the
design of the fiber interconnect.

The permanent splice, unlike the connector, is not
designed for repeated coupling and uncoupling.   As a result,

splices generally have lower losses and are easier and simpler to fabricate than connector couplings. Connectors require precision mating hardware attached to the fiber, making them more complicated and costly. Since fiber losses as low as 0.2 dB/km can be achieved, it is clear that the splice loss required in long-distance transmission systems must be held to a minimum. This requires tight tolerances on fiber parameters, along with precise mechanical alignment and proper fiber end preparation. The splice must be able to be performed under field conditions by personnel who are not highly trained. This dictates that the technique be performable in minimum time with minimum fiber handling. The resulting joints must be mechanically rugged, insensitive to environment, and have a suitably long life.

Techniques have been developed to splice both two single fibers and multifiber arrays. Single-fiber splicing techniques are appropriate for use on cables containing 10 or fewer fibers, whereas multifiber techniques must be used on cables containing large numbers of fibers. Single-fiber splice techniques include snug-fitting tubes, precision-aligned grooves, and the thermal fusion.

In the snug-fitting tube approach, fibers are inserted into the opposite ends of the bore of a tube fabricated from either glass or metal, which may or may not contain index-matching fluid. One such design (3.12), (shown in Figure 3.17), uses a snug-fitting glass tube tapered at the ends to ease insertion of the fiber. Once the fibers are in place, index-matching epoxy is inserted through the hole located in the middle of the tube. Mean-field splice losses of 0.21 dB have been demonstrated with this approach.

Another snug-fitting tube approach has recently been reported (3.13) that compensates somewhat for variations in fiber outside diameter. The tube is made from an elastomeric polyester.

When the fibers are inserted into the bore, the elastic material exerts forces on the fibers, which automatically align their axes, even if the fibers have different outside diameters. Splices that use index-matching fluid in the elastic tube have average insertion losses of 0.15 dB.

Another splice approach is the "loose-tube" splice (3.14). Although the fibers are aligned by inserting them into a capillary tube, this splicing technique is really an example of the precision aligned-groove approach where the fibers are forced to lie in a precision groove. The loose tube consists of a square capillary; the interior corner forms the precision groove. The fibers are bent upon inserting them into the tube, forcing them to line in one of the interior corners. Mean splice losses of 0.1 dB have been reported with this approach. An illustration of such a splice is shown in Figure 3.18.

Simple fiber splicing can also be achieved by thermally fusing the two fibers that are to be joined. Low-loss fusion splices have been accomplished using a surrounding Nichrom heater (3.15), electric-arc (3.16), a microtorch (3.17), or a CO_2 laser (3.18). Losses of a few tenths of a decibel are obtainable by thermal fusion.

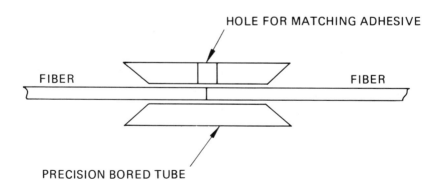

Figure 3.17. Snug-fitting glass tube splice (3.12).

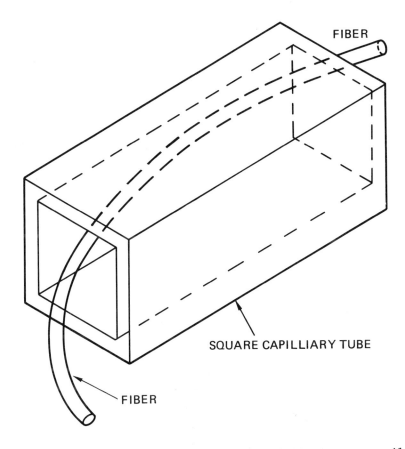

FIBER

SQUARE CAPILLIARY TUBE

FIBER

Figure 3-18. "Loose tube" splice design employing square capillary whose interior corner forms a precision groove
(ref. 3.14).

Multifiber cable-splicing techniques have been developed using precision aligned grooves to hold the individual fibers in multifiber arrays. Preferentially etched silicon chips (3.19) can be used to fabricate arrays which, on the average, deviate from perfect uniformity by only 2.5 µm. An array formed by interleaving etched silicon chips and layers of fibers is shown in Figure 3.19. The assembly is formed by removing all coating material from the linear arrays (ribbons) of fibers and interleaving the fibers with etched silicon alignment chips as illustrated in the figure. Once the stack is formed, it is potted in a potting material. Fiber end surfaces are prepared by grinding and polishing the whole array. Splicing is achieved by butting together two such formed arrays using unoccupied grooves on the top and bottom chips for array alignment. Arrays containing 144 fibers (12 12-fiber ribbons) can be joined in this fashion.

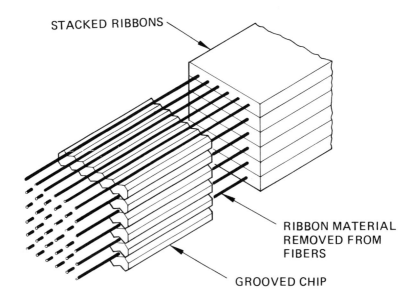

Figure 3.19.   Two-dimensional array formed by interleaving etched silicon chips and layers of fibers (3.19).

The evaluation of cable splices containing such large numbers of fibers would be tedious and time consuming if each fiber joint were characterized individually. A "figure of merit," which characterizes the degree of uniformity in an array, provides a basis for alignment comparison of connector halves without requiring complete splices to be assembled and losses measured. The figure of merit of a splice connector half is obtained by measuring the location of each fiber in the array, then transforming these data by rotation and translation into a coordinate system used to define a perfectly uniform array. The magnitude of the vector $\bar{q}_k$ joining the kth point of the uniform array to the corresponding k^{th} point in transformed measured data arrays is computed for each point. A calculation is then performed which alters the parameters of the uniform array until a "best fit" to the measured data is obtained. This occurs when the mean value of $|\bar{q}_k|$ is a minimum. The minimum value is defined as the figure of merit for the array. Using preferentially etched silicon chip arrays with figures of merit of 2.5 μm or less have been reported (3.19).

Techniques for fabricating connectors include snug-fitting tubes, precision grooves, and lensed systems. Two snug-fitting tube approaches are the metal bushing, split-alignment tube design (used by Hughes Aircraft Company) and the injection-molded-plastic biconical connector (developed by Bell Telephone Laboratories). A schematic illustration of the metal bushing, split-alignment tube design is shown in Figure 3.20. The biconical design is schematically shown in cross section in Figure 3.21. Both designs of demountable connectors result in insertion losses of less then 1 dB.

A design using a precision groove, or cusp, when four glass rods are fused to form a "loose alignment tube," has been developed by TRW cinch. The alignment tube is bent at either end to ensure that when the fibers to be coupled enter the ends of

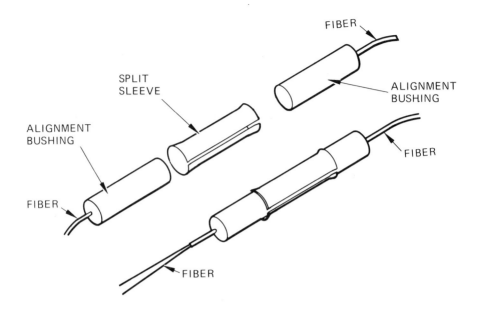

Figure 3.20.   Schematic illustration of metal bushing, split align-
                ment tube design developed by Hughes Aircraft
                Company.

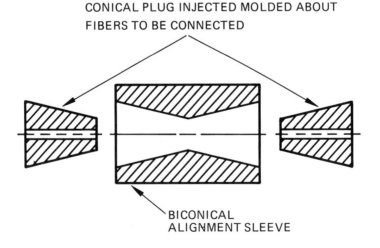

Figure 3.21.   Schematic illustration of biconical design developed
                Bell Laboratories.

the tube they automatically find their way into the same vee-shaped interstice. A schematic illustration of this connector along with the connector shell is shown in Figure 3.22. This design also results in insertion losses of less than 1 dB.

The demountable connector designs mentioned are but a few of many currently either available or under investigation. Most electrical connector manufacturers also offer a fiber connector. The interested reader is referred to the various manufacturer's specification sheets and catalogs for further information.

TRW CINCH OPTALIGN CONNECTOR

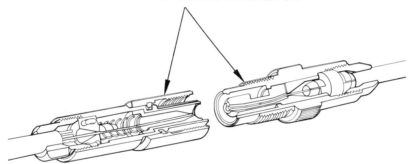

Figure 3.22.   Schematic illustration of "loose alignment tube" design developed by TRW CINCH.

3.4   SUMMARY

In this chapter an elementary treatment of some of the engineering design considerations that must be considered in the design of input coupling components for fiber optic transmission lines has been presented. A brief discussion of splice and demountable coupling is also given.

PROBLEMS FOR CHAPTER 3

1.   Calculate the efficiency of coupling between an LED
        with a 55 µm active diameter and an isotropic emission
        pattern and a step index fiber with a 0.2 NA and 55 µm
        core diameter. What would the coupling efficiency be
        if the fiber were graded? Repeat the calculations
        assuming a Lambertian emission pattern. Are the
        numbers calculated the maximum achievable?

2.   A surface-emitting LED with an 18 MIL active emitting
        area diameter emits a total of 10 mW of optical power
        at maximum rated drive current. A similar diode with
        a 2 MIL active diameter emits 1 mW of optical power.
        For each diode, find the maximum amount of power that
        can be coupled into a graded index fiber with an axial
        NA of 0.2 and a 55-µm-diameter core.

3.   Design a taper launcher to maximize the coupling from
        a 25 µm diameter surface-emitting LED into a 55 µm core
        diameter graded fiber with NA = 0.2. What is the maxi-
        mum achievable coupling efficiency?

4.   What would you expect the maximum theoretical coupling
        efficiency to be between a single-mode HeNe laser with a
        1 mm beam diameter and a single-mode fiber with a 5 µm
        core diameter?

REFERENCES

3.1   C.A. Burrus and B.I. Miller, *Opt. Commun. 4*, 307 (1971).

3.2   Refer to Chapter 4.

3.3   Robert D. Maurer, "Introduction to Optical Waveguide
      Fibers," in *Introduction to Integrated Optics*, M.K. Barnoski,
      Ed. (Plenum, New York, 1974), Chapter 8.

3.4   Colin Pask and Allan W. Snyder, *Opto-Electronics 6*, 297
      (1974).

3.5   K.H. Yang and J.D. Kingsley, *Appl. Opt. 14*, 268 (1975).

3.6   M. Born and E. Wolf, *Principles of Optics* (Pergamon Press,
      Oxford, 1965), p. 189.

3.7   M.C. Hudson, *Appl. Opt. 13*, 1029 (1974).

3.8   T. Ozeki and B.S. Kawasaki, *Electron. Lett. 12*, 607 (1976).

3.9   Y. Uematsu and T. Ozeki, "Efficient Power Coupling Between
      a MH LED and a Multimode Fiber with Tapered Launcher,"
      *Tech. Digest of 1977 Intern. Conf. Integrated Optics and
      Optical Fiber Commun.*, Tokyo, Japan (1977), p. 371.

3.10 S. Horiuchi, K. Ikeda, T. Tanaka, and W. Sasuki, *IEEE Trans.
      ED-24*, 986 (1977).

3.11 T.C. Chu and A.R. McCormick, *Bell Syst. Tech. J. 57*, 595
      (1978).

3.12 H. Murata et al., "Splicing of Optical-Fiber Cable on Site,"
      *Proc. First European Conference on Optical Fiber Communi-
      cation*, London, Sept. 1975.

3.13 Mark L. Oakss, W. John Carlsen, and John E. Benasutti,
      "Field-Installable Connectors and Splice for Glass Optical
      Fiber Communications Systems," 12th Annual Connector Symp.,
      Cherry Hill, New Jersey, Oct. 17, 1979.

3.14 C.M. Miller, *Bell Syst. Tech. J. 54*, 1215 (1975).

3.15 D.L. Bisbee, *Bell Syst. Tech. J. 50*, 3153 (1971).

3.16   Y. Kohanzadeh, *Appl. Opt. 15*, 793 (1976).

3.17   R. Jocteur and A. Tardy, *2nd Proc. European Conf. on Optical Fiber Communications*, 261 (1976).

3.18   K. Egashira and M. Kogayashi, *Appl. Opt. 16*, 1636 (1977).

3.19   C.M. Miller, *Bell Syst. Tech. J. 57*, 75 (1978).

CHAPTER 4

ELECTROLUMINESCENT SOURCES FOR FIBER SYSTEMS

H. Kressel

RCA Laboratories
Princeton, New Jersey 08540

4.1   INTRODUCTION

The laser diodes and light-emitting diodes (LEDs) of
primary interest in fiber systems emit in the 0.8 to 0.9 and
1.3 μm spectral regions where the transmission losses are minimal.
Source power and modulation requirements depend on the system
objective as well as the type of fibers used, and the various
possible options will not be discussed here. It is desirable,
however, to develop a light source with the widest possible system
utility, although its full potential would usually not be used.
For this reason, there has been a longstanding interest in a laser
diode for optical communications which combines high radiance,
ease of direct modulation to gigahertz rates, small size, and low
cost. However, it is only since the development of the AlGaAs/
GaAs heterojunction laser structures in 1968 (4.1-4.3), and important technological progress made after that time, that the potential of laser diodes for fiber communications is being realized.

In parallel with laser diode development, it was recognized that specially designed LEDs could be useful with relatively
short fiber lengths. Light-emitting diodes have less restrictive
operating conditions than those required for lasers. (For example,

the temperature dependence of the power output from a laser diode
is much greater than from an LED.) The LED requirements for
fiber communications have called for the development of special
structures capable of reliable high-current density operation,
a high modulation rate, and a choice of materials to meet the
above spectral requirements. Laser diodes and LEDs share many
aspects of a common technology, and both have benefited from
the reliability advances made in recent years.

Literature concerning all aspects of laser diodes and
electroluminescent devices is available (4.4), and the present
review is necessarily limited. An introductory treatment concerning spontaneous and stimulated emission in Section 4.2 is followed
in Section 4.3 by a discussion of laser diode operation and the
role of heterojunctions. Section 4.4 is concerned with the
material requirements and the role of interfacial defects. State-
of-the-art cw laser diode operation is discussed in Section 4.5,
including structural and modal properties. In Section 4.6 the
requirements for the LEDs and the major structures used are discussed. Section 4.7 is concerned with lasers and LEDs for 1.0 to
1.7 µm emission. Major factors affecting diode reliability are
reviewed in Section 4.8. Finally, Section 4.9 reviews laser diode
modulation.

4.2   SPONTANEOUS AND STIMULATED EMISSION

4.2.1   Spontaneous Emission

The electroluminescent diode is basically a device in
which electrons and holes are injected into the p- and n-type
regions, respectively, by the application of a forward bias V, as
shown in Figure 4.1. The current-voltage characteristics are
described by the well-known diode equation

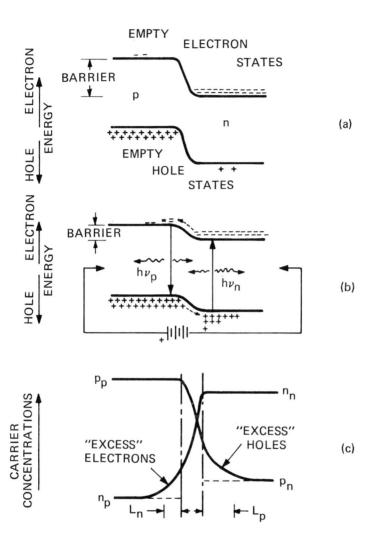

Figure 4.1. Electroluminescent p-n junction operation. (a) Zero
bias. The built-in potential drop across the p-n junction represents a large barrier for the motion of the
electrons and holes. (b) In forward-bias, the potential barrier is significantly reduced by the application of the external voltage. (c) Majority and
minority carrier concentrations on the n- and p-sides
of a forward-biased p-n junction.

$$I = I_s \ [\exp(qV/akT) - 1] \quad , \qquad (4.1)$$

where I_s is the saturation current and $1 < a < 2$. The injected
minority carriers recombine either radiatively or nonradiatively.
If they recombine radiatively, the emitted photon energy, $h\nu$,
is approximately equal to the bandgap energy E_g (neglecting
recombination via deep centers). For nonradiative recombination,
the energy released is dissipated in the form of heat. The two
processes are characterized by minority carrier lifetime, τ_r
and τ_{nr}, for radiative and nonradiative recombination, respectively. These determine the internal quantum efficiency, η_i, and
the lifetime, τ,

$$\eta_i = \left(1 + \frac{\tau_r}{\tau_{nr}} \right)^{-1} \qquad (4.2)$$

$$\frac{1}{\tau} \cong \frac{1}{\tau_r} + \frac{1}{\tau_{nr}} + \frac{2S}{d} \quad , \qquad (4.3)$$

where S is the interfacial recombination velocity in a double
heterojunction diode with heterojunction spacing d.

In the simplest model, the nonradiative lifetime decreases
linearly with the density of nonradiative recombination centers,
N_t, their capture cross section, σ_t, and the electron thermal
velocity, v_{th},

$$\tau_{nr} \approx (N_t \sigma_t v_{th})^{-1} \quad . \qquad (4.4)$$

The nature of the nonradiative centers is still not well understood in GaAs and related compounds. However, it is generally
believed that lattice defects such as vacancies, interstitials,
states associated with dislocations, and precipitates of commonly

used dopants form nonradiative recombination sites. It is essential, therefore, that the materials used for light-emitting devices be as free from defects as possible.

The reason for the selection of certain semiconductors for LEDs and laser diodes is the relationship between the radiative lifetime and the bandstructure. In "direct bandgap" semiconductors, electrons and holes across the gap from each other have the same value of crystal momentum, making possible direct recombination without phonon participation to conserve momentum; this process is very efficient, and the resulting lifetime is very short (10^{-10} to 10^{-8} sec). The group of direct-bandgap semiconductors includes, among others, GaAs, InAs, InP, $Al_x Ga_{1-x} As$ ($0 \leq x \leq 0.4$) and $GaAs_{1-y}P_y$ ($0 \leq y \leq 0.45$). On the other hand, in indirect bandgap semiconductors, such as Si, Ge, GaP, and AlAs, the participation of phonons is required to conserve momentum in across-the-gap electron-hole recombination, and the resultant recombination process is relatively slow.

From the measured absorption coefficient of a semiconductor as a function of photon energy, it is possible to calculate the radiative recombination coefficient, B_r, which determines the radiative minority carrier lifetime for a given majority carrier concentration, n_o and p_o, respectively:

$$\tau_r \cong (B_r n_o)^{-1} \qquad \text{(n-type material)}$$

$$\tau_r \cong (B_r p_o)^{-1} \qquad \text{(p-type material)} \quad . \qquad (4.5)$$

The above expressions assume that the injected carrier density is substantially below the majority carrier density. If this is not the case, then we enter a regime of bimolecular recombination and the above expressions overestimate the lifetime (see Eq. (4.17)).

Table 4.1 shows the vastly different theoretical values
of B_r for direct and indirect energy bandgap materials (4.5).
In Si, for example, B_r = 1.79 x 10^{-15} cm^3/sec, whereas in GaAs,
B_r = 7.21 x 10^{-10} cm^3/sec. For a majority carrier density P_o =
10^{17} cm^{-3}, the calculated radiative lifetime in Si is τ_r = 6 x
10^{-3} sec; in GaAs τ_r = 1.4 x 10^{-8} sec. The consequent effect
on the internal quantum efficiency becomes obvious when we consider some reasonable values for the nonradiative lifetime.
Assuming even a low density of nonradiative centers, e.g.,
10^{15} cm^{-3}, with σ_t = 10^{-15} cm^2, we calculate $\tau_{nr} \approx 10^{-7}$ sec.
From Eq. (4.2) the calculated internal quantum efficiency there-
fore would be only 1.7 x 10^{-5} in Si, but 0.88 in GaAs. A high
internal quantum efficiency is clearly much easier to achieve
in direct bandgap than in indirect bandgap semiconductors.

Table 4.1. Calculated Recombination Coefficient for Representative
Direct- and Indirect-Bandgap Semiconductors (Ref. 4.5)

Material	Energy Bandgap Type	Recombination Coefficient, B_r (cm^3/sec)
Si	Indirect	1.79 x 10^{-15}
Ge	Indirect	5.25 x 10^{-14}
GaP	Indirect	5.37 x 10^{-14}
GaAs	Direct	7.21 x 10^{-10}
GaSb	Direct	2.39 x 10^{-10}
InAs	Direct	8.5 x 10^{-11}
InSb	Direct	4.58 x 10^{-11}

4.2.2   Stimulated Emission

Stimulated emission is achieved by carrier population
inversion, a condition where the upper of two electronic levels
separated in energy by $E = E_2 - E_1$ has a higher probability of
being occupied by an electron than the lower level. The probability of a photon (with energy $h\nu \approx E$) inducing a downward (induced)
electron transition will then exceed the probability for an upward
transition, i.e., photon absorption. Light amplification becomes
possible, therefore, when an incident photon stimulates the emission of a second photon with energy approximately equal to the
energy separation of the electronic levels.

The above concepts are general, but semiconductor lasers
differ in detail from gas or other types of solid-state lasers
where the radiative transitions occur between discrete levels
of spatially isolated excited atoms. The spontaneous radiation
produced by transitions between the isolated atoms extends over
a very narrow spectral range, whereas in semiconductors the active
atoms are very closely packed causing their energy levels to
overlap into bands. The high packing density (about 10^{18} cm^{-3})
of excited atoms in a semiconductor, compared with only about
10^{10} cm^{-3} in a gas laser, is advantageous because the optical
gain coefficient is relatively high, thus allowing much shorter
optical cavities.

Population inversion in a semiconductor is illustrated
in Figure 4.2, which shows the electron energy as a function
of the density of states in an undoped semiconductor at a temperature sufficiently low (T = 0 K in the illustration) for the
conduction band to be empty of electrons. When electrons are
injected, they fill the lower energy states of the conduction
band to F_c, the quasi-Fermi level for electrons. An equal density
of holes is generated to conserve charge neutrality in the material. The states in the valence band to F_v are, therefore, empty
of electrons. Photons with energy greater than E_g, but less

than $F_c - F_v$, cannot be absorbed (since the conduction band states
are occupied), but these photons can induce downward electron
transitions from the filled conduction band states into the empty
valence band states. With increasing temperature, electrons
and holes are redistributed, thus smearing out the sharply defined
carrier distributions shown in Figure 4.2. However, the basic
conditions for stimulated emission remain defined as above in
terms of the separation of the quasi-Fermi levels, $F_c - F_v < h\nu$.

For stimulated emission to occur, the gain must match
the optical losses at some photon energy within the spontaneous
radiation spectrum. Since the photon density is highest at or

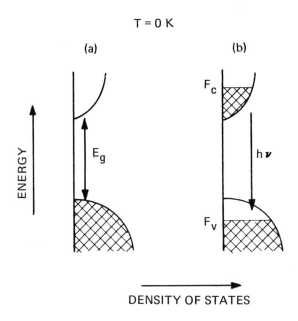

Figure 4.2.   Electron energy as a function of the density of
states in an intrinsic direct bandgap semiconductor
at T = 0 K.   (a) Equilibrium and   (b) under high
injection.

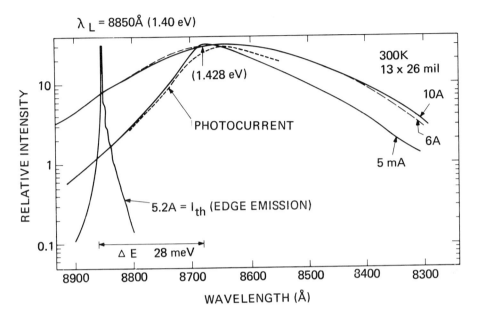

Figure 4.3.   Spontaneous spectra at 300 K below and above
lasing threshold as viewed through the surface
of a diode. The lasing spectrum was observed
by viewing the edge of the diode. The photocurrent curve shows the dependence of the diode
photocurrent on the incident photon wavelength.

near the peak of the spontaneous emission, this is where the
gain coefficient peaks.   Figure 4.3 shows the position of the
lasing peak at threshold with respect to the spontaneous emission spectrum in a laser diode where both can be observed with

no distortion resulting from internal absorption (4.6). The
lasing peak energy is below the spontaneous emission peak by
$\sim$0.02 to 0.03 eV at room temperature, for lightly doped material,
with the separation decreasing in value with temperature as shown
in Figure 4.4.

If the semiconductor is highly doped ($\sim$10^{18} cm^{-3} in GaAs),
the radiative transitions may involve impurity states. In the
case of p-type GaAs (acceptor ionization energy $\sim$0.03 eV), the
emitted photon energy is below the bandgap energy. For n-type
GaAs, where donors have typical ionization energies <0.01 eV,
the Fermi level shifts upward with doping into the conduction
band, and the photon energy exceeds the nominal bandgap energy.
If the material is heavily doped and compensated, the lasing photon energy is below the bandgap energy. The lasing wavelength
can be varied from 0.85 to 0.95 µm in GaAs at room temperature
with doping variation, but the device performance is usually best
in the 0.88 to 0.91 µm range.

4.3   STRUCTURAL REQUIREMENTS FOR EFFICIENT LASER DIODE
        OPERATION

A unique feature of the laser diode is the ability to
obtain stimulated emission by minority carrier injection using
a p-n junction or heterojunction. The efficient operation of a
laser diode requires effective carrier and radiation confinement
to the vicinity of the junction. The average injected electron
density from an n-type region into a p-type region (under conditions where the reverse process of hole injection is negligible)
is

$$N_e = J\tau/ed \quad , \tag{4.6}$$

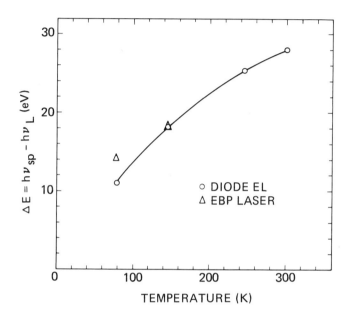

Figure 4.4.   Energy separation between the diode spontaneous emis-
               sion band peak $h\nu_{sp}$ and lasing peak at threshold $h\nu_L$
               between 77 and 300 K. Also shown for comparison are
               two data points for an electron-beam-pumped laser
               (EBP) made from GaAs similar to that in active region
               of heterojunction laser diode (Ref 4.6).

where e is the electron charge, J is the injection current density
and d is the effective active region width in which the average
injected electron density is N_e with lifetime τ.

In a typical GaAs laser diode the excess carrier pair
density needed to reach lasing threshold is 1 to 2 x 10^{18} cm^{-3}
at room temperature. To minimize the threshold current density,
we restrict the width of the recombination region by placing a
potential barrier for minority carriers a distance less than the
diffusion length from the injecting p-n junction. Figure 4.5
shows, for example, how a p-p heterojunction presents a potential
barrier for electrons if at room temperature ΔE_g > 0.1 eV. This
is the single heterojunction, "close-confinement" laser (4.1, 4.2).
The use of an injecting p-n heterojunction yields the double
heterojunction diode (4.3). It is, however, essential that the
heterojunction interface be relatively defect-free to prevent
excessive nonradiative recombination of the injected carriers
(see Section 4.4).

To ensure wave propagation in the plane of the junction,
means must be provided for at least partial stimulated radiation
confinement to the region of inverted population. An effective
way of providing the required dielectric profile for radiation
confinement is by the use of heterojunctions. Two heterojunctions,
as indicated in Figure 4.6, provide a controlled degree of radiation confinement because of the higher refractive index in the
lower bandgap recombination region. The fraction of the radiation confined depends on the heterojunction spacing d and on the
refractive index steps at the lasing wavelength.[*] In general,
an equal refractive index step Δn at each heterojunction is used
to prevent the loss of waveguiding that can occur in thin asym-
metrical waveguides. However, even in the symmetrical double

----

[*] In $Al_x Ga_{1-x} As$/GaAs structure, $\Delta n \cong 0.62x$ at $\lambda \cong 0.9$ μm.

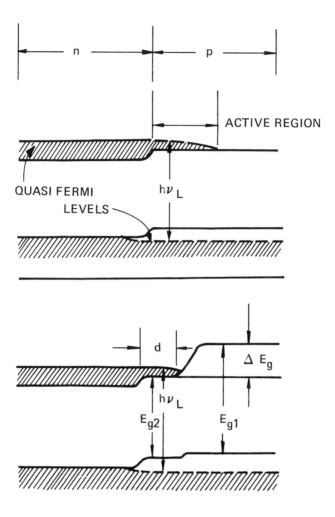

Figure 4.5.   Electron distribution in a forward-biased homojunction without a potential barrier for carrier confinement (top), and with a p-p heterojunction (bottom)
placed at a distance d (less than the diffusion length)
from the p-n junction.

heterojunction laser, wave confinement within d is eventually
reduced when the heterojunction spacing becomes too small (4.7).
This effect is illustrated in Figure 4.6 for a relatively wide
double heterojunction structure in which the radiation confinement is nearly complete (Figure 4.6(c)), and for a very narrow
heterojunction spacing in which the optical field intensity
spreads symmetrically on the two sides (Figure 4.6(e)). A controlled degree of radiation spread is used to obtain devices having a desired far-field radiation pattern.

The fraction of the radiation confined to the recombination region of the double heterojunction laser, denoted Γ, affects
the radiation pattern and threshold current density. The far-
field pattern is affected because of the change in effective source
size. The threshold is affected because only the fraction of the
optical power within the recombination region is amplified.

Optical feedback is obtained in a laser diode by cleaving
two parallel facets to form the mirrors of the Fabry-Perot cavity.
The lateral sides of the laser are either formed by roughening
the device edges by wire sawing to form a "broad-area" diode,
Figure 4.7(a), or by confining the ohmic contact to a selected
area to produce a stripe-contact diode, Figure 4.7(b). This will
be discussed further in Section 4.5.

The conditions for gain in a semiconductor laser cavity
have been treated in detail elsewhere (4.4) and will not be considered here. It has been found by many researchers that the
threshold current density of typical double heterojunction AlGaAs-
GaAs lasers decrease linearly with d (4.4(f),(h)).

$$\frac{J_{th}}{d} = (4.0 \pm 0.5) \times 10^3 \text{ A cm}^{-2} \text{ }\mu m^{-1} , \qquad (4.7)$$

when $0.3 \leq d \leq 2$ μm. Hence, for $d = 0.3$ μm, $J_{th} \cong 1200$ A/cm^2,
a satisfactory value for reliable room temperature cw operation.

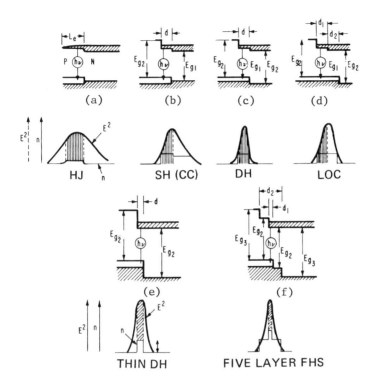

Figure 4.6.   Energy diagram distribution of the refractive index
(n) and of the optical energy (E^2), and position of
the recombination region in major laser diode structures. (a) Homojunction; (b) single heterojunction (close confinement) diode; (c) double heterojunction diode with essentially full carrier and
radiation confinement; (d) large optical cavity
(LOC) diode; (e) double heterojunction diode with
full carrier confinement, but only partial radiation
confinement; (f) four-heterojunction diode.

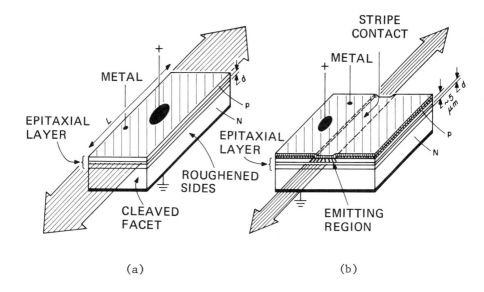

(a)   (b)

Figure 4.7. (a) Broad-area and (b) stripe-contact laser diodes.

In addition to the threshold current density, the differential quantum efficiency above threshold, η_{ext}, and the overall power conversion efficiency, η_p, are important laser diode parameters. In state-of-the-art pulsed-operation laser diodes, η_{ext} = 40 to 50% at room temperature (emission from both sides). The power conversion efficiency peaks at 2 to 4 times I_{th}, and the best values at room temperature are ∿22% for two-sided emission, but the power conversion efficiency is only a few percent near threshold.

The double heterojunction laser diode is currently most widely used for cw operation, but more complex structures are being studied for special applications. Figure 4.6(d) shows the schematic configuration of the large optical cavity (LOC) (4.8) laser diode in which the p-n junction is bracketed by two heterojunctions, and the four-heterojunction diode (4.9) where the outer two heterojunctions are added to the two inner heterojunctions to obtain additional control of the optical cavity width. A discussion of these structures is beyond the scope of the present review (4.4(f)).

4.4   MATERIALS

Table 4.2 shows the III-V compounds with direct bandgap energies of interest in the 0.8 to 0.9 and 1.0 to 1.7 μm spectral ranges. Both vapor-phase epitaxy (VPE) (4.10) and liquid-phase epitaxy (LPE) are commonly used to prepare the materials shown (4.11). The Al-containing alloys such as AlGaAs are most frequently prepared by LPE because of chemical problems associated with the vapor deposition of these materials. Where there is a significant lattice parameter mismatch between the epitaxial layer and the substrate, VPE is advantageous because of the ease of compositional grading to reduce the defect density in the active region of the grown layer. For example, GaAsP is grown

Table 4.2   III-V Materials for 0.8 to 0.9 μm and ~1.0 to 1.7 μm Emission

Wavelength (μm)	E_g (eV)	Material	Substrate	$\Delta a_o/a_o$ (%)[a]
~1.0 to 1.1	1.2	$In_{0.2}Ga_{0.8}As$	GaAs	1.38
		$GaAs_{0.85}Sb_{0.15}$	GaAs	1.13
1.0 to 1.7	0.73 to 1.24	$In_xGa_{1-x}As_xP_{1-y}$	InP	~0[b]
~0.88 to 0.91	~1.42	GaAs	GaAs	0
~0.82	~1.55	$Al_{0.12}Ga_{0.88}As$	GaAs	0.017
		$GaAs_{0.86}P_{0.14}$	GaAs	0.50

[a] Lattice parameter mismatch between epitaxial layer and substrate.

[b] The lattice matching condition is $y = 2.16(1-x)$.

exclusively by VPE on GaAs or GaP substrates, whereas InGaAsP
can be grown by both techniques on InP.

The choice of materials for laser diodes is limited by
the need to incorporate heterojunctions for carrier and radiation
confinement, as discussed above. Although LEDs do not require
heterojunctions, they are desirable for efficient optical communications sources (Section 4.6).

The major problem associated with heterojunction structures arises from the interfacial lattice parameter mismatch.
Consider joining two hypothetical simple cubic lattice materials
as shown in Figure 4.8. A dislocation network is generated to
accommodate the lattice parameter misfit. A linear dislocation
density is theoretically formed to fully relieve the strain,

$$\rho_\ell = \frac{\Delta a_o}{a_o^2} ,$$

where Δa_o is the small difference in lattice parameter.

The actual misfit dislocation density depends on the
crystal structure and on the elastic strain in the material; i.e.,
fewer dislocations are generated if the crystal remains strained
to partially accommodate the lattice misfit. All of the generated dislocations do not simply lie in the interface; some are
inclined and propagate into the grown layer. If the average
"inclined" dislocation spacing is less than a minority carrier
diffusion length in the vicinity of the injected minority carriers, then the effective diffusion length, and hence the minority carrier lifetime, is reduced.

The immediate question of interest here is the effect
of interfacial dislocations on the recombination velocity. Consider the (100) plane in the sphalerite structure. Assuming that
each state associated with the dislocation core constitutes a
nonradiative center, the calculated (4.12) interfacial surface
recombination velocity S is

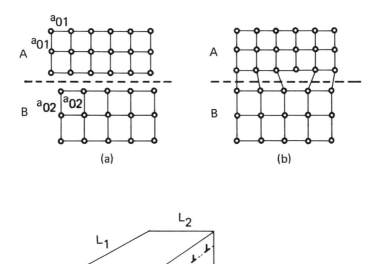

$$L_d \approx \bar{a}_0^2 / \Delta a_0$$

Figure 4.8.   Schematics showing the formation of an edge misfit
              dislocation in joining simple cubic crystal A with
              lattice constant a_{01} and substrate B with lattice
              constant a_{02}.   (a) Separate crystals;   (b) formation
              of edge dislocation when crystals are joined;
              (c) formation of dislocations at edge of crystal L_1
              x L_2.   The distance between dislocations is L_d.

$$S \cong \frac{8v_{th}\sigma_t}{a_o^2} \left(\frac{\Delta a_o}{a_o}\right) \text{ cm/sec} \quad , \qquad (4.8)$$

where v_{th} is the thermal carrier velocity and σ_t is a capture cross section. Assuming $\sigma_t = 10^{-15}$ cm^2 and $a_o \approx 5.6$ Å, Eq. (4.8) predicts

$$S \cong 2.6 \times 10^7 \; (\Delta a_o/a_o) \quad . \qquad (4.9)$$

The experimental values indeed fall in the range of

$$S = (3.8 \pm 1.2) \times 10^7 \; (\Delta a_o/a_o) \qquad (4.10)$$

in $In_yGa_{1-y}P/GaAs$ heterojunction structures where y was varied to change the lattice misfit (4.13). For comparison, note that $S \cong 10^6 - 10^7$ cm/sec for an unpassivated GaAs surface.

It is evident from Eq. (4.3) that a high value of S can drastically reduce the carrier lifetime of heterojunction diodes (4.14). If $\tau_{nr} \gg \tau_r$, we can express the internal quantum efficiency for spontaneous recombination in the approximate form,

$$\eta_i \cong (1 + 2S\tau_r/d)^{-1} \quad . \qquad (4.11)$$

For example, if $\tau_r = 2.5 \times 10^{-9}$ sec and d = 0.5 μm, then a value of $\eta_i \geq 50\%$ requires that $S \leq 10^4$ cm/sec.

The AlAs-GaAs alloy system meets the above criterion, since the lattice parameter difference between GaAs and AlAs is $\Delta a_o/a_o = 0.14\%$, with smaller differences, of course, for reduced Al concentration differences at the heterojunctions. Concerning S in GaAs-AlGaAs heterojunctions, available data indicate that $S < 5 \times 10^3$ cm/sec, a value which is satisfactory for efficient narrow recombination region devices (4.15).

It is possible, however, to match lattice parameters
at heterojunction structures by a judicious choice of binary,
ternary, and quaternary materials (4.16). Figure 4.9 shows how
lattice matching combinations can be fabricated. For the longer
wavelength devices of interest in optical communications, lattice-
matching InGaAsP/InP structures are desirable, as will be discussed in detail later.

4.5   LASER DIODES OF $Al_xGa_{1-x}As$ FOR ROOM TEMPERATURE CW
        OPERATION

4.5.1   Structures

Double heterojunction AlGaAs lasers easily operate cw
at room temperature with up to 12% Al concentration in the recombination region (λ_L = 0.9 to 0.8 μm); the typical cw threshold
current density is below 2000 A/cm^2, and devices with J_{th} = 1000
to 1500 A/cm^2 are routinely used.

The broad area laser structure formed by sawing the edges
perpendicular to the cleaved facets forming the mirror cavity,
Figure 4.7(a), has been widely used for pulsed power operation
and in early double heterojunction laser studies (4.3, 4.17, 4.18),
but cw laser diodes now use a stripe-contact geometry exclusively
for several reasons: (1) The radiation is emitted from a small
region, which simplifies coupling of the radiation into fibers
with low numerical aperture. (2) The operating current can be
under 0.1 A because it is relatively simple to form a small active
area with convenient large area contacting procedures. (3) The
thermal dissipation of the diode is improved compared with a mesa
diode because the heat-generating active region is embedded in
an inactive semiconductor medium. (4) The small active diode area
makes it simpler to obtain a reasonably defect-free area.

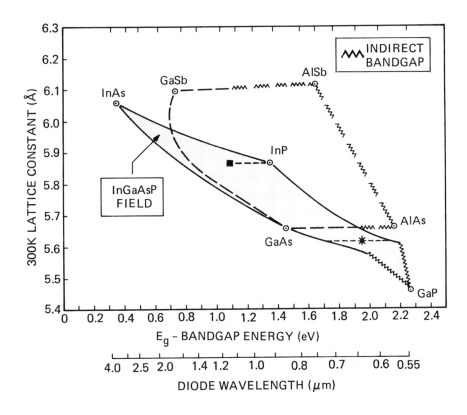

Figure 4.9.   Lattice parameter a_0 and bandgap energy variation
with composition for major III-V compounds.

(5) The active region is isolated from an open surface along its
two major dimensions, a factor important for reliable long-term
operation (Section 4.8).

The simplest structures to fabricate are the planar types
using one of three basic methods for defining the active area.
In the first (4.19), Figure 4.10(a), isolation is obtained by
opening a stripe contact (typically between 8 and 60 μm wide) in
a deposited SiO_2 film. The surface of the diode is then metal-
lized with the ohmic contact formed only in the open area of the
surface. A second method of stripe formation (4.20) uses selec-
tive Zn diffusion through an n-type region at the surface to
reach the p-type layers underneath, Figure 4.10(b). The third
method (4.21) uses proton bombardment to form a resistive region
on either side of the desired recombination region, Figure 4.10(e).

In all of these structures, radiation and current spreading occurs on either side of the stripe, depending on the resistance and thickness of the layers between the surface and the
recombination region, and on the diffusion length in the recombination region (4.22). Figure 4.11(a) shows a typical cross section of a laser diode made using oxide stripe isolation (4.23)
with 10% Al in the recombination region, designed for emission at
∿8200 Å. An example of the near field in the plane of the junction is shown in Figure 4.11(b). Although the radiation intensity
is highest in the central region of the 12 μm wide stripe, there
is significant radiation extending to about 20 μm (4.24).

Diodes designed for cw operation are generally mounted
p-side down on copper heat sinks to minimize the thermal resistance. A soft solder such as indium is commonly used to minimize
strains in the devices.

4.5.2   Electrical and Optical Properties

The threshold current density of heterojunction lasers
typically increases with temperature T as $\exp(T/T_o)$, where

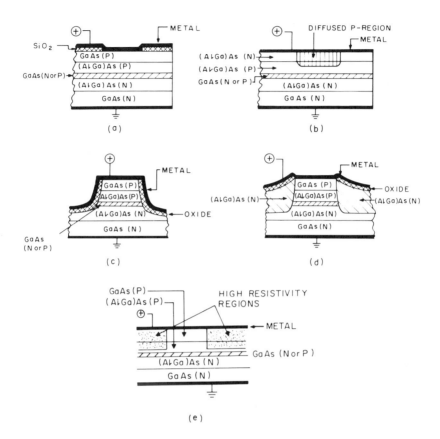

Figure 4.10.   Relatively simple stripe-contact diode structures.

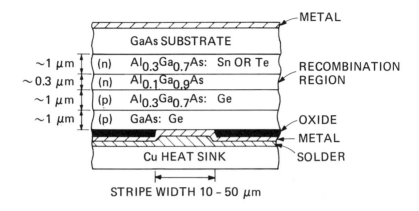

(a)

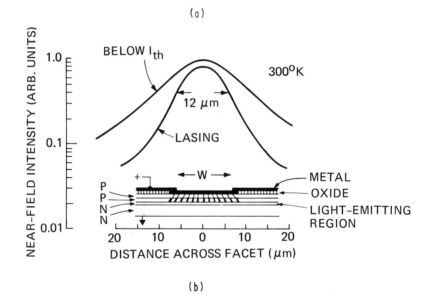

(b)

Figure 4.11   (a) Schematic laser cross section (not to scale) and
               (b) near-field pattern in the plane of the junction
               (Ref. 4.24).

$T_o \approx$ 50–200 K, while the lasing peak energy decreases with
increasing temperature at a rate of $\sim 5 \times 10^{-4}$ eV K^{-1}. It is evi-
dently desirable to produce devices with as low a temperature
dependence of J_{th} as possible. Within limits, the larger the
bandgap energy step at the heterojunctions the lower the temperature dependence of J_{th}, i.e., the larger T_o is. In AlGaAs structures a value of $T_o \cong$ 200 K is achieved with bandgap steps of
about 0.4 eV.

The maximum output from a pulsed laser occurs at a duty
cycle determined by its thermal and electrical resistance, the
threshold current and T_o value. The electrical resistance of
a 13 μm wide stripe laser is generally between 0.4 and 1 Ω, and
the thermal resistance is 12 to 20 K/W. When the power dissipation caused by the diode electrical resistance is small compared
with the power dissipation in the junction itself, the criterion
for cw operation is (4.25),

$$\frac{I_{th} E_g R_{th}}{eT_o} < 0.37 , \qquad (4.12)$$

where R_{th} is the thermal resistance, e is the electron charge, and
T_o is the parameter in the temperature dependence of J_{th}.

For illustration, consider a laser diode with $R_{th} =$
14 K/W, $E_g =$ 1.5 eV, and $T_o =$ 50 K. From Eq. (4.12), a threshold
current as large as 0.88 A could permit cw operation; however,
in practice, lower threshold values are required because of the
impact of the electrical resistance at high current levels.

The lateral width of the emitting region can be adjusted
for a desired operating level, and 100 mW of peak cw power (one
facet) is obtainable for stripe widths of 100 μm (4.24). However, for the typical power levels needed in optical communications (3 to 10 mW), stripe widths of 6 to 20 μm are generally
used. These stripe widths provide a useful compromise between
low operating currents and reasonable power emission level. Further reduction in the stripe width in planar structures does not

yield a corresponding threshold current reduction because of the
imperfect lateral current and optical confinement.

Threshold current and power output as a function of diode
current and temperature is shown in Figure 4.12 for a diode with
a 12 μm stripe. The junction temperature of such devices is a
few degrees above the heat sink temperature.

The lowest threshold current devices are the mesa and the
buried heterojunction lasers, Figure 4.10(d), if the emitting
region is 1 to 2 μm wide (4.26). The cw power emitted by such
narrow devices is typically limited by reliability considerations
to the 1 to 2 mW range. However, cw threshold currents as low as
10 mA have been achieved at room temperature with buried heterojunction lasers.

4.5.3   Radiation Pattern and Modal Properties

The laser modes are separable into two independent sets
with transverse electric (TE) and transverse magnetic (TM) polarization. The modes of each set are characterized by mode numbers
m, s, and q along the three axes of the cavity, transverse, lateral
and longitudinal, respectively.

The *longitudinal* mode spacings depend on the index of
refraction and its dispersion, and on the cavity length. The
spectral emission lines are spaced a few angstrom units apart in
typical lasers.

The *lateral* modes are dependent on the method used to
define the two edges of the laser and on the refractive index profile which is affected by the current and gain distribution in the
junction plane.

The *transverse* modes (direction perpendicular to the
junction plane) depend on the dielectric profile perpendicular to
the junction plane. In the devices discussed here, only the fundamental mode is excited. This is achieved by restricting the
thickness of the waveguiding region (i.e., heterojunction spacing) to values well under one micrometer. Therefore, the

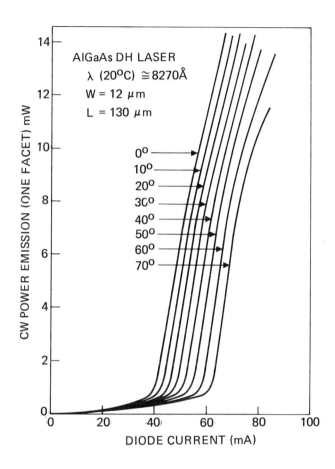

Figure 4.12. (a) Power output as a function of current and heat sink temperature of AlGaAs cw laser.

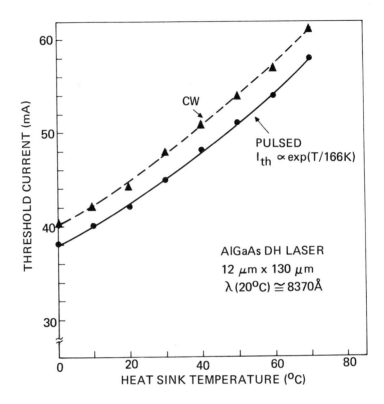

Figure 4.12.   (b)   Comparison of pulsed and cw operation threshold
current as a function of heat sink temperature.

far-field consists of a single lobe in the direction perpendicular
to the junction. Higher order transverse anodes give rise to
"rabbit-ear" lobes which are undesirable for fiber coupling.

For a laser operating in the fundamental *transverse* mode,
the far-field full angular width at half power, $\theta_\perp$, is a function
of the near-field. Figure 4.13 shows $\theta_\perp$ calculated as a function
of d and Δn in double heterojunction lasers (4.4(f)). In practical cw laser diodes, $\theta_\perp$ = 35 to 50° with d = 0.2 to 0.3 µm, consistent with Δn = 0.12 to 0.18. The beam width in the direction
parallel to the junction plane is typically 6 to 10° and varies
with the diode topology and internal geometry. An example of
the far-field pattern of a typical cw laser diode is shown in
Figure 4.14. From one-third to one-half of the power emitted from
one facet can be coupled into a step index fiber with a numerical
aperture (NA) of 0.14.

While fundamental transverse mode operation is easily
achieved, most laser diodes operate with several longitudinal
modes with a spectral width of 10 to 15 Å. However, some lasers
operate predominantly in a single longitudinal mode, at least
in some current range, as shown in Figure 4.15. Very short cavity
lasers ($\sim$100 µm long) and lasers specially constructed for fundamental lateral mode operation (see below) frequently exhibit above
average spectral purity.

Lasers operating in the fundamental *lateral* mode can
be produced by restricting the lateral width of the active region.
The simplest case to analyze occurs when there are two high (and
equal) steps Δn perpendicular to the junction plane forming a
rectangular box cavity. Consider a diode of width W. From a
simple analysis of the critical angle for total internal reflection at the sidewalls, we find that the highest lateral mode number S_m (denoting the fundamental mode as 1) capable of propagating
in the structure is

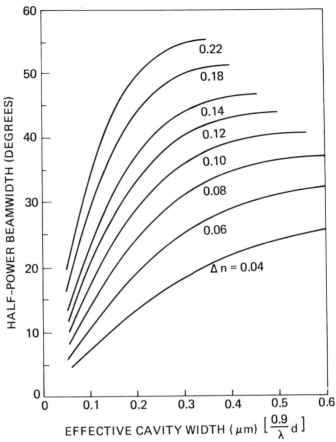

Figure 4.13.   Calculated beam width in the direction perpendicular
               to the junction plane for double heterojunction
               diodes as a function of the cavity width d and
               refractive index step Δn (Ref. 4.4f).

$$S_m = \text{Integer} \left[1 + \left(\frac{2nW}{\lambda} \right) \left(\frac{2\Delta n}{n} \right)^{1/2} \right] . \qquad (4.13)$$

Therefore, to operate in the fundamental lateral mode, the index
step condition is

$$\frac{\Delta n}{n} \lesssim \frac{1}{8} \left(\frac{\lambda}{nW} \right)^2 . \qquad (4.14)$$

For GaAs, $n = 3.6$ and $\lambda \approx 0.9$; hence $\Delta n/n \lesssim 7.8 \times 10^{-3} \, W^{-2}$, where
W is in micrometers.

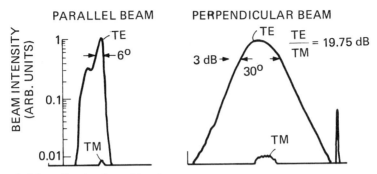

Figure 4.14.   Typical radiation patterns in the plane of the junc-
                tion and perpendicular to the junction for a
                laser diode of the type described in Ref. 4.23.

The above step-index model is not appropriate for planar
stripe lasers because the shallow dielectric profile is related
to the current and gain distribution; hence, it is subject to
change with current and optical power level. Experimentally,
fundamental lateral mode operation (at least near threshold) is
frequently obtained in planar stripe lasers with stripe widths
$\lesssim 10$ μm, but higher-order modes generally reach threshold with
increasing current, producing a multilobed near and, hence, far-
field pattern.

Mode guiding in planar stripe lasers results from several
contributions to the index profile:

- Increasing gain near the stripe center
  increases the imaginary part of the
  dielectric constant

- Local heating, related to the current distribution and power dissipation, increases
  the real part of the dielectric constant

- The higher carrier density, which lowers
  the index, acts in the opposite direction
  to reduce the dielectric constant under the
  stripe.

A shallow maximum in the index under the stripe contact
results from the combined effect of the above three factors. However, changes in current and optical power density affect the
ability of the various lateral modes to propagate.

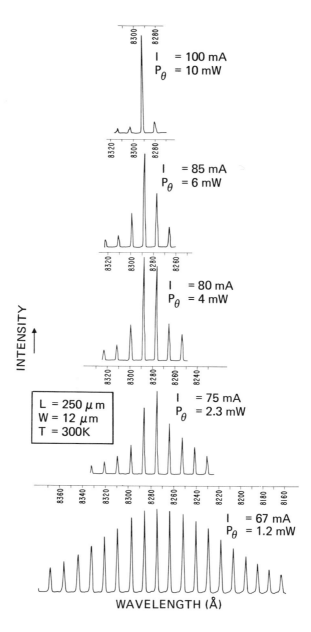

Figure 4.15. CW lasing spectrum of an AlGaAs DH laser as a function of current.

Fundamental lateral mode operation is desirable because it eases coupling into low numerical aperture fibers (including single-mode fibers) and because mode changes with current are frequently accompanied by "kinks" in the power output versus current curves.

The key principle that governs the restriction of laser operation to a single mode is that the difference between the propagation losses of the fundamental and the higher-order modes be as large as possible. Restricting the stripe width of lasers to very small values is one method of achieving this objective, although at the expense of the useful power from the device. Another approach consists of incorporating regions in the device that produce a greater internal absorption coefficient for the higher-order modes than for the fundamental one.

Forming the stripe contact in other configurations such as misaligning it with respect to the mirrors (4.27), curving it (4.28), or bending it (4.29), as well as altering its width (4.30) along the length have all been employed with some success, but at the sacrifice of the threshold current density. Furthermore, since the mode stability depends on the current flow, changes in temperature, duty cycle, and threshold alter the spatial pattern of these devices.

The most successful structures for providing lateral mode stability incorporate a lateral waveguide. Two approaches are used. In one class of structures, the lateral mode is stabilized by introducing loss into the wings of the mode so that higher order modes, which distribute more power to the wings, have higher thresholds. In the other class of structures, a lateral index of refraction profile is provided so that a waveguide is formed that only supports the fundamental mode. In Figure 4.16 we illustrate some of these laterally guided structures.

The simplest structure is the buried heterojunction (BH) structure (4.26), Figure 4.16(a), where the lateral guide is formed

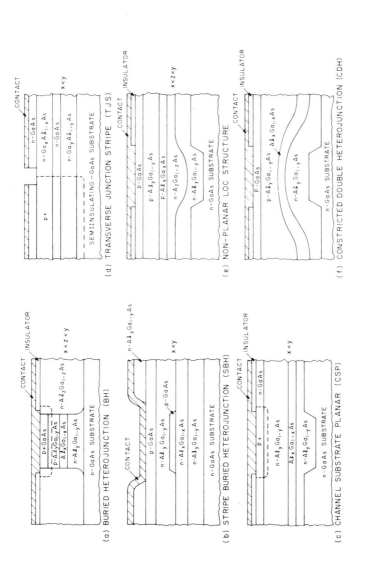

Figure 4.16. Schematic cross section of some of the laser configurations designed for fundamental lateral mode operation. The shaded areas are post-growth Zn diffusions to form heavily-doped, p-type regions which help confine the injected carriers.

by etching a mesa and regrowing a lower index material. However, the diode must be very narrow (1 to 2 μm) for fundamental mode operation, thus limiting the power to 1 to 2 mW. To improve the power output, the structure can also be a large optical cavity (BH-LOC) (4.31). In a similar LOC type structure shown in Figure 4.16(b), called a stripe-buried-heterostructure (SBH) (4.32), only the active region is etched and the waveguide is regrown around it while the remainder of the optical cavity remains untouched. In the channel-substrate-planar device (4.33) Figure 4.16(c), absorption in the wings is used. The transverse junction stripe laser (TJS) (4.34), Figure 4.16(d), employs an injecting homojunction in the lateral direction, while in the transverse direction there is an isotype double heterojunction structure.

A number of devices other than the CSP structure are grown on etched well substrates (4.35). For example, a LOC structure can be grown in a well (4.36), Figure 4.16(e), on the well shoulder (4.37), or between two wells as in the double-dovetail confined double heterojunction laser (4.38) shown in Figure 4.16(f). All rely on the nonplanarity in the active (i.e., light-guiding) region to provide the lateral waveguide.

The lasers that operate in the fundamental lateral mode also have linear power output versus current characteristics. In addition, they frequently exhibit single longitudinal mode operation over a useful current range, but the dominant mode shifts with temperature.

4.6   LIGHT-EMITTING DIODES

The typical spectral bandwidth of a light-emitting diode (LED) is typically 300 to 400 Å at room temperature, at least one order of magnitude broader than the laser diode emission. Because of the increased wavelength dispersion in the fiber, this breadth limits the bandwidth for long-distance fiber communications. Furthermore, the coupling efficiency into low numerical aperture fibers is much lower than for laser diodes. However, the LED has the advantage of a simpler construction and a smaller temperature dependence of the emitted power. For example, the output from a double heterojunction (DH) LED typically decreases by only a factor of two as the heat sink temperature increases from room temperature to 100°C (at constant current).

Heterojunction structures are desirable for high-radiance, high-speed LEDs because it is possible to surround an appropriately doped recombination region with high-bandgap energy material having low internal absorption of the emitted radiation. Furthermore, waveguiding leads to enhanced edge emission, resulting in a more directional beam than otherwise achieved from a surface-emitting LED.

4.6.1 Diode Topology

The LED topology is designed to minimize internal reabsorption of the radiation, allow high current density operation (greater than 1000 A/cm^2), and maximize the coupling efficiency into fibers. Although the structures described here are applicable to all materials, most of the reported work has been on GaAs, AlGaAs, and InGaAsP diodes.

Two basic diode configurations for optical communications have been reported: surface emitters (4.39) and edge emitters (4.40, 4.41). In the most efficient surface emitter the recombination region is placed close to a heat sink, as shown in Figure 4.17, and a well is etched in the GaAs substrate to accommodate a fiber. The emission from such a diode is Lambertian.

The edge-emitting heterojunction structures use the partial internal waveguiding of the spontaneous radiation to obtain improved directionality of the emitter power in the direction

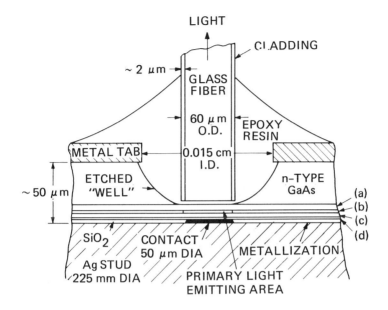

Figure 4.17.   Cross-sectional drawing (not to scale) of small-area
double heterojunction electroluminescent diode
coupled to optical fiber. Layer (a), n-type
$Al_xGa_{1-x}As$, 10 μm thick; emitting layer
(b), $p-Al_yGa_{1-x}As$, ≈1 μm, thick; layer (c), p-type
$Al_xGa_{1-x}As$, 1 μm thick; layer (d), p-GaAs, ≈0.5 μm
thick (for contact purposes) (Ref. 4.39).

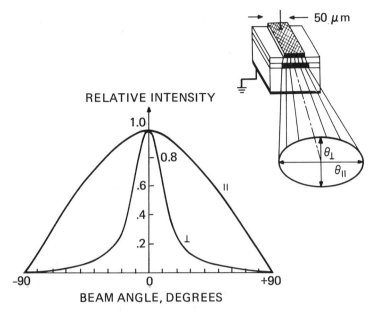

Figure 4.18.   Spontaneous emission pattern from edge–emitting LED
designed for low beam divergence in the direction
perpendicular to the junction plane.

perpendicular to the junction plane.   This is illustrated in

Figure 4.18 (4.42).   In edge-emitting structures it is desirable

to deposit an antireflecting film on the emitting facet to increase

the efficiency.   The lateral width of the emitting region is

adjusted for the fiber dimension, but it is typically 50 to 100 μm.

Surface-emitting and edge-emitting LEDs emit several

milliwatts of power at current densities in excess of 1000 A/cm^2,

with the maximum current being limited by junction heating.   Com-

parative data for several LED structures are shown in Table 4.3.

The coupling efficiency of surface emitters into fibers is typi-

cally lower than that from edge emitters.   Figure 4.19 shows plots

of the coupling efficiency from various device types into step-

index fibers as a function of the numerical aperture.   The cou-

pling efficiency into graded index fibers is substantially less

for equal numerical apertures.

Finally, we note some measurements of noise from GaAs

homojunction and heterojunction LEDs.   Above 5 kHz, the homojunc-

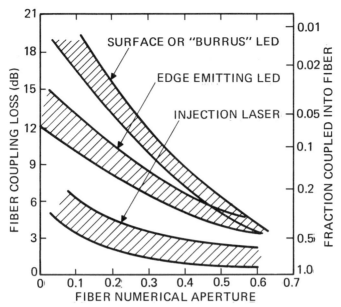

Figure 4.19. Calculated coupling efficiency into step index fibers of varying numerical aperture. The shading indicates the range of values expected (M. Ettenberg, unpublished).

tion LED was reported to be an ideal radiation source (4.43). In the case of heterojunction structures, with good ohmic contacts, the measured 1/f noise would be well below the shot noise in a practical system (4.44).

4.6.2   Modulation Capabilities

With the diode always in forward bias, the dependence of the power output from an LED on modulation frequency, ω, is given by (4.45),

$$\frac{P(\omega)}{P_o} = \frac{1}{[1 + (\omega\tau)^2]^{1/2}} ,\qquad (4.15)$$

where τ is the minority carrier lifetime in the recombination region, and P_o is the dc power emission value. (However, parasitic circuit elements can reduce the modulated power range below that indicated in Eq. (4.15).)

Table 4.3.　Comparative LED Data

Material	Wavelength (μm)	Construction	Radiance (W/cm²-Sr) at [Current] (mA)	Highest Emitted Power (mW)	Quantum Efficiency (%)	Maximum Modulation Bandwidth (MHz)	Optical Power Into Fiber[c] (mW)
GaAs[1] (HOMO)	0.9	S.E.	25[300]	—	~1	450[a]	—
InGaAs[1] (HOMO)	1.06	S.E.	15[100]	—	~1	150[a]	—
InGaAsP/InP[2] (DH)	1.2	S.E.	50[100]	~6	3	90[a]	—
AlGaAs[3] (DH)	0.85	S.E.	100[150]	14	7.6	10[b]	0.21
AlGaAs[4] (DH)	0.85	E.E.	1000[450]	8	~1	200[a]	0.85
AlGaAs[5] (DH)	0.82	S.E.	200[300]	15	~3	17[a]	—

Code:　HOMO:　Homojunction
　　　　DH:　Double-Heterojunction
　　　　S.E.　Surface-Emitter with etched well (Burrus type)
　　　　E.E.　Edge-Emitter

[a]3 dB reduction in optical power　　[b]6 dB reduction in optical power
[c]N.A. = 0.15, step index fiber with diameter of 80 μm.

(1)　R.C. Goodfellow and A.W. Mabbit, Electron. Lett. 12, 51 (1976).

(2)　A.G. Dentai, T.P.Lee, C.A. Burrus, and E. Buehler, Electron. Lett. 13, 484 (1977).

(3)　F.D. King, J. Straus, D.I. Szentesi, and A.J. SpringThorpe, Proc. IEE 126, 619 (1976).

(4)　M. Ettenberg, H. Kressel, and J.P. Wittke, IEEE J. Quantum Electron. 12, 360 (1976).

(5)　T.P. Lee and A.G. Dentai, IEEE J. Quantum Electron. 14 150 (1978)

The electrical power at the detector is proportional to the square of the optical power. The bandwidth capability f_c in terms of the electrical power is defined at the level where

$$P^2(\omega) = \frac{1}{2} P_o^2 .$$

Hence, from Eq. (4.15),

$$f_c = \frac{1}{2\pi\tau} . \tag{4.16}$$

We can relate f_c to diode parameters as follows. The radiative carrier lifetime is a function of the carrier density. Hence, it changes with increasing current density J, as we now show:

$$\tau_r = \frac{ed}{2J} (p_o + n_o) \left\{ [1 + 4J/eB_r d (p_o + n_o)^2]^{1/2} - 1 \right\} . \tag{4.17}$$

Here, p_o and n_o are the electron and hole concentrations in the recombination region of thickness d without injection, e is the electron charge, and B_r is the recombination coefficient which varies with the semiconductor (Table 4.1).

At low injection levels in a p-type recombination region assuming $\tau_r \ll \tau_{nr}$,

$$f_c \cong (B_r p_o / 2\pi) , \tag{4.18}$$

whereas at high injection levels

$$f_c \cong \frac{1}{2\pi} \left(\frac{B_r J}{ed} \right)^{1/2} . \tag{4.19}$$

It is evident that a high-speed diode requires the lowest possible value of τ but without sacrifice in the internal quantum efficiency.

In GaAs and related compounds, a high density of nonradiative centers are formed when the dopant concentration approaches the solubility limit at the growth temperature. Of the devices reported so far, Ge-doped DH LEDs have exhibited outstanding modulation capability (200 MHz) and moderate efficiency. The use of Ge is advantageous because it can be incorporated into GaAs to levels of $\sim10^{19}$ cm^{-3} (yielding minority carrier lifetime values of 1 to 2 nsec) without catastrophically reducing the internal quantum efficiency (4.46). Other dopants, including Zn and Si, have been used to fabricate slower diodes. Figure 4.20 shows experimental and calculated plots of $P(\omega)/P_o$ for heterojunction diodes having different values of minority carrier lifetime obtained by varying the dopants in the recombination region. The data for these broad-area diodes are in agreement with Eq. (4.15), indicating that no other modulation-rate limiting factors were present in those devices. (These devices were operated in a current range where high injection effects were not dominant.)

4.7   LASERS AND LEDs FOR 1.0 to 1.7 MICRON EMISSION

Lasers and LEDs emitting at wavelengths in excess of 1 μm at room temperature can be produced using InGaAs, InGaP, GaAsSb or InGaAsP in the recombination region. Heterojunction structures of InGaAsP/InP dominate because there is no lattice parameter mismatch (4.47). The design considerations of double heterojunction InGaAsP/InP lasers are similar to those of AlGaAs/GaAs lasers. Recombination region thickness values between 0.2 and 0.5 μm are used to obtain the lasing threshold current densities below 3000 A/cm^2 needed for convenient cw operation at room temperature. Figure 4.21 shows the cross section of a typical DH laser diode using an oxide isolated stripe contact. Figure 4.22 shows the power output as a function of current at various heat sink temperatures (4.48).

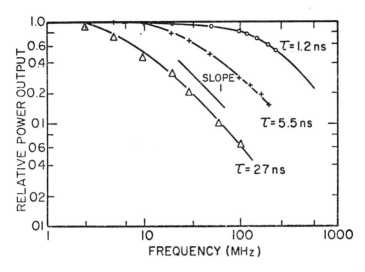

Figure 4.20.   Relative power output as a function of the modulation
frequency for diodes having different minority car-
rier lifetimes.

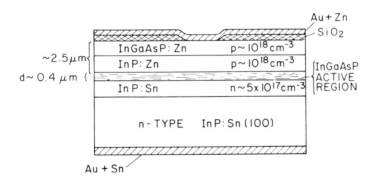

Figure 4.21.   Cross section of DH InGaAsP/InP laser.

InGaAsP/InP LEDs are also constructed using double
heterojunction structures and both surface and edge-emitting
devices have been produced. Figure 4.23 shows the power output
as a function of current at various temperatures of an edge-
emitting LED. The modulation properties of these devices are
similar to those of AlGaAs DH LEDs, i.e., modulation above
100 MHz can be obtained as indicated by carrier lifetimes of
~2 nsec (4.49).

4.8   DIODE RELIABILITY

The two basic failure modes that may limit the operating life of electroluminescent devices involve facet damage and
internal defect generation ("gradual" degradation). The first
depends on the optical flux density and the pulse length and is
relevant only to laser diodes, whereas the second depends on the
current density, the duty cycle, and the quality of the device.
Incoherent emitters are affected by the gradual degradation phenomena and share with laser diodes common properties in this
regard. Table 4.4 summarizes the major degradation modes.

Table 4.4. Relationship of Degradation Cause to Operation

Large facet damage ◄────►	Optical power density, pulse length
Facet erosion ◄────►	Optical power density, ambient
Internal defects[a] ◄────	Current density ($J^{1.5-2}$), temperature
Contact degradation[a] ◄────►	Current density, temperature

[a]Degradation rate similar in LED and laser operation.

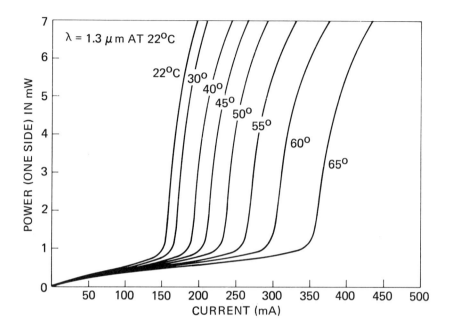

Figure 4.22. CW power output as a function of current at various heat sink temperatures (G.H. Olsen, unpublished).

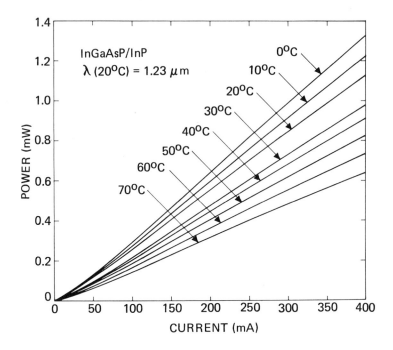

Figure 4.23.   Emitted power from InGaAsP/InP LED (edge-emitting)
as a function of heat sink temperature.

4.8.1   Facet Damage

Facet damage may be subdivided into two categories. Catastrophic degradation refers to large-scale damage which typically occurs quite rapidly. Facet erosion is a relatively modest damage level that has a lesser impact on the facet reflectivity and hence on the device performance. Facet erosion typically develops slowly in the course of operation.

Facet damage caused by intense optical fields is a well-known phenomenon in solid-state lasers and occurs in semiconductor lasers of all types under varying conditions. The appearance of the damaged laser in the catastrophic mode suggests local dissociation of the material at the facets. A commonly used figure of merit is the critical damage level observed in watts per centimeter of the emitting facet, P_c. For a given operating condition, P_c decreases linearly with decreasing emitting region width. For example, in double heterojunction lasers operated with 100 nsec pulses, $P_c \cong 200$ W/cm when d = 1 μm, with P_c increasing to about 400 W/cm when d = 2 μm. It is important to note, however, that these figures represent only average values for broad-area diodes because damage is commonly initiated at portions of the facet that contain mechanical flaws.

The critical damage level is also affected by the applied pulse length, with P_c decreasing with pulse length, t, as $t^{-1/2}$ (4.50), at least for pulse lengths between 20 and 2000 nsec. It is not surprising, therefore, that catastrophic damage can occur in laser diodes operating in the cw range at their maximum emission levels, even with the relatively low-power densities achievable. Because of the nonuniform radiation distribution in the plane of the junction in stripe-contact lasers, it is difficult to establish simple linear power density criteria. However, the damage threshold for 100 nsec pulse length operation

is about 10 times higher than in cw operation of diodes selected
from the same group (4.24). The damage is generally initiated in
the center of the diode in the region of highest optical intensity at a power level of 2 to 3 mW/μm (2 to 4.2 x 10^5 W/cm^2 considering the radiation distribution in the direction perpendicular to the junction plane).

Facet damage can be prevented by operation at reduced
power levels and by the use of dielectric facet coatings. The
coatings reduce the electric field at the surface by reducing the
reflectivity. The experimental data (4.51) fit the following
expression (4.52),

$$\frac{P'_c}{P_c} = \frac{n(1 - R')}{(1 + R'^{1/2})^2} \quad . \quad\quad (4.20)$$

R' must be chosen to give the desired facet protection without
unduly increasing the device threshold current.

Facet erosion may occur even when the laser is emitting
power levels well below those needed for catastrophic degradation.
Operation in a moist ambient accelerates the facet damage effect
possibly because of oxidation accelerated by the optical power at
the laser facet (4.53). It has been conclusively shown that facet
erosion is prevented by the use of Al_2O_3 facet coatings, thus
extending the life of lasers under the usual operating conditions
(4.54). The Al_2O_3 film is half-wave thick and, therefore, the
facet reflectivity is not changed. As a result the threshold current of the laser is unchanged by the addition of the film.

4.8.2   Gradual Degradation

Gradual degradation in an incoherent emitter results in
a reduction in the externally measured quantum efficiency. In
a laser diode, gradual degradation means that the threshold

current density increases and the differential quantum efficiency
decreases, resulting in a reduced power output for a given current
density without evidence of facet damage. Of course, if a laser
is operated near threshold, it may cease lasing altogether for
a small increase in J_{th}. However, in this case a small increase
in current can usually recover the initial power value. Therefore, the definition of degradation in a laser diode is somewhat
arbitrary since current adjustments can keep a laser operating
despite changes in its efficiency with time.

Gradual degradation can occur in both homojunction and
heterojunction lasers (4.55). The available evidence indicates
that the internal quantum efficiency is reduced by the formation
of nonradiative recombination centers within the recombination
region (4.56) and that the internal absorption coefficient can
increase. It is well known that the minority carrier lifetime
decreases as degradation progresses, suggestive of a decrease in
τ_{nr}. This is consistent with the observed reduction in the external quantum efficiency of diodes operating in the spontaneous
emission mode. Whether lasing or not, diodes of a given construction degrade similarly at a given current density. The presence
of a p-n junction is not needed for the occurrence of gradual
degradation since optically excited GaAs shows evidence of nonradiative center formation in the region where electron-hole pair
recombination occurs (4.57).

Defect formation is commonly spatially nonuniform, and
it is dependent on the type of semiconductor, the degree of per-
fection of the material, and the method used to fabricate and
assemble the diode. For example, dislocations and exposed active
region edges should be minimized (4.23), and strains introduced
in the process of assembling diodes should be avoided. The use
of $Al_xGa_{1-x}As$ ($x \approx 0.05 - 0.1$) in the recombination region is
desirable to improve diode life (4.58).

The available evidence suggests that the gradual
degradation process is initiated by flaws initially present in
the recombination region of the diode. These grow in size as
a result of point defects which could indiffuse from areas adjacent to the active region. One prominent effect in some degraded
lasers is the formation of "dark lines" in which the luminescence
is greatly reduced (4.59). These regions have been identified
as large dislocation networks which, starting at existing smaller
dislocation networks, grow by the aggregation of vacancies or
interstitials (4.60). In addition, more dispersed nonradiative
centers such as native point defects could contribute to the
degradation process.

The origin of these point defects is still the subject
of speculation. Gold and Weisberg (4.61), in their GaAs tunnel
diode degradation studies, suggested that nonradiative electron-
hole recombination at an impurity center could result in its displacement into an interstitial position, leaving a vacancy
behind. It is assumed that multiphonon emission gives an intense
vibration of the recombination center which effectively reduces
its displacement energy. Applying this mechanism to the electroluminescent efficiency degradation, the assumption is that the
vacancy and interstitial formed have a large cross section for
nonradiative recombination. Repeated nonradiative recombination
gradually moves it to internal sinks.

The following hypothesis is a working explanation for
the major degradation effects. In essence, the "phonon-kick"
process uses part of the energy released in nonradiative recombination to enhance the diffusion of vacancies of interstitials.
(Whether any point defects are actually formed within the recombination region remains unclear.) Hence, if nonradiative
electron-hole recombination occurs at the damaged surface of a
diode (as in the case of the sawed-edge diode experiment described
in Ref. 4.23), it accelerates the motion of point defects into the
active region of the device. Thus, the stoichiometry of the

material in or in close proximity to the active region can affect the local density of defects (4.62). Regions where nonradiative recombination occurs will tend to grow in size, leading to the nonuniform degradation process commonly observed.

From the available results, control of gradual degradation appears to be primarily a metallurgical problem with many of the contributing factors known and others still to be defined.

The progress made to date suggests that the operating lifetimes for both LEDs and cw laser diodes are in the range of practical fiber systems requirements. AlGaAs LEDs with half-lives in excess of 50,000 hours are a reality. Continuous wave laser diodes have also operated for periods of time in excess of 50,000 hours, although current adjustments are needed in most cases to maintain a power output of a few milliwatts.

Data show that the diode degradation rate increases with operating temperature (4.63), which provides a method for accelerating life tests.

A review of the available data (4.64) for AlGaAs cw laser diodes suggests that the useful life of laser diodes is reduced with temperature following an expression of the form $\exp(E/kT)$ where E is an effective "activation energy" which may vary from 0.6 to about 0.9 eV. A value of about 0.7 eV appears to fit the data well. Furthermore, the aging data follow a log-normal distribution with regard to failures, although, as shown in Figure 4.24, there are deviations for long operating times.

Figure 4.25 shows room temperature operating data for a group of 24 cw AlGaAs lasers. These data, which show that the failures follow the log-normal curve, predict a mean time to failure in the vicinity of 100,000 hours. Lifetimes in excess of these values should be possible with improving technology.

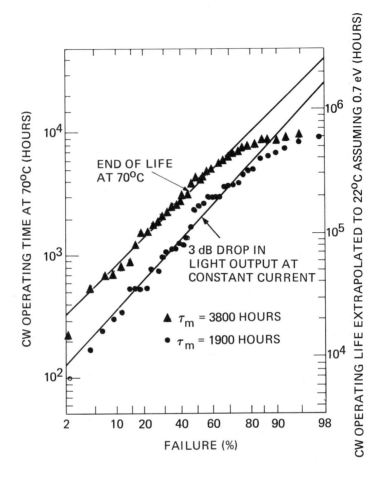

Figure 4.24.   Time to failure at 70°C heat sink temperature on
               log-normal coordinates for 40 low threshold (∿50 mA)
               oxide-defined stripe AlGaAs DH lasers. Two failure
               points are assumed, the first when the laser drops
               to half its initial output at constant current
               (3 dB life) and the second where the laser can no
               longer emit 1.25 mW at the 70°C heat sink tempera-
               ture (end of life). τ_m is the time need for 50% of
               the lasers to fail. The estimated lifetime at 22°
               is shown assuming 0.7 eV "activation" energy.
               (Ref. 4.64).

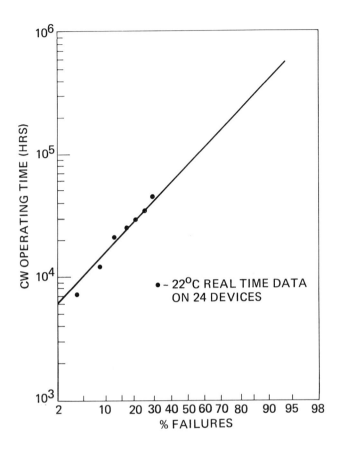

Figure 4.25.   Time to failure on log–normal coordinates for 24
AlGaAs DH lasers emitting ∿10 mW cw at 22°C heat
sink temperature. Failure is defined as the inabil-
ity of the laser to emit 5 mW cw. Test and device
details are discussed in Ref. 4.64.

4.9   LASER DIODE MODULATION AND TRANSIENT EFFECTS

The ability to modulate the laser diode output by current changes is a major advantage of the semiconductor laser over
other laser types, but a number of complicated factors determine
the useful modulation rate (4.65, 4.66). In this section we
review what happens when a laser is turned on with a step current, and modulation rate limitations when operating in the
lasing region.

The essential difference between the modulation of an
LED and a laser diode arises from the difference between the spontaneous and stimulated carrier lifetime. The spontaneous carrier lifetime (see Eq. (4.17)), which ultimately determines the
rate at which the LED can be modulated, is a function of the carrier concentration in the recombination region and of the injection level, but it is independent of the photon density in the
device.

In a laser diode, on the other hand, the recombination
is a *stimulated* process above threshold and the carrier lifetime
is thereby reduced. In a single-mode laser, the carrier lifetime
decreases above threshold starting from the 2 to 5 nsec spontaneous recombination lifetime at threshold. Therefore, it becomes
possible to modulate a semiconductor laser at GHz rates, but only
if the device is prebiased at or above threshold. As we discuss
in Section 4.9.1, prebiasing is also essential to eliminate the
delay between the current and the light pulse.

4.9.1   The Laser Turn-On Delay

In this section we discuss the delay between the current and the stimulated light emission from a laser which is
turned on with a step-current pulse.

When a step current is applied to a diode initially
off, the injected carrier pair density N_e within the recombination

region d wide increases at a rate,

$$\frac{dN_e}{dt} = J/ed - N_e/\tau \quad , \tag{4.21}$$

where τ represents the average lifetime of the carriers in the recombination region. The delay time t_d needed to reach the threshold carrier pair density N_{th} is from Eq. (4.21),

$$t_d \cong \int_0^{N_{th}} \frac{dN_e}{J/ed - N_e/\tau} \quad . \tag{4.22}$$

Hence,

$$t_d \cong \tau \ell n \, [I/(I - I_{th})] \quad . \tag{4.23}$$

The delay can be reduced by a bias current $I_b < I_{th}$,

$$t_d \cong \tau \ell n \, [I/(I - (I_{th} - I_b))] \, . \tag{4.24}$$

The delay disappears when the bias current $I_b = I_{th}$. High-frequency modulation is normally performed with prebias, but prebiasing is not needed for moderate modulation frequencies. Suppose, for example, that we require the power emitted at a current $I = 2I_{th}$. From Eq. (4.23), the delay is $t_d \cong 0.69 \, \tau$, or typically 1.4 to 3.5 nsec. Thus, by using current pulse lengths of at least about 10 nsec, the laser output can be comfortably modulated in the low megahertz range without a prebias current.

4.9.2   Oscillatory Effects

Oscillatory effects in lasers are associated with the coupling between the carrier population and the photon density in the cavity. A theoretical treatment of the known effects is beyond the scope of this chapter, but insight into the most

important of these effects can be gained from the greatly
simplified model below.

Assume that the laser oscillates in a single mode whose
excitation level can be described by a (uniform) photon density
N_{ph}. The corresponding excess carrier pair density in the recombination region, N_e, is also assumed to be spatially uniform.
The output coupling and internal losses combine to give a photon
lifetime τ_{ph}, and the (spontaneous) carrier lifetime τ is assumed
to be constant. The temporal variations of the populations are
given by two nonlinear, coupled equations:

$$\frac{dN_e}{dt} = \underbrace{\frac{J}{ed}}_{\substack{\text{carrier} \\ \text{injection}}} - \underbrace{AN_e N_{ph}}_{\substack{\text{carrier decrease} \\ \text{from stimulated} \\ \text{emission}}} - \underbrace{\frac{N_e}{\tau}}_{\substack{\text{decrease from} \\ \text{spontaneous} \\ \text{recombination}}}$$

$$(4.25a)$$

$$\frac{dN_{ph}}{dt} = \underbrace{AN_e N_{ph}}_{\substack{\text{increase by} \\ \text{stimulated} \\ \text{photon} \\ \text{emission}}} - \underbrace{\frac{N_{ph}}{\tau_{ph}}}_{\substack{\text{loss of} \\ \text{photons}}} + \underbrace{\frac{\gamma_s N_e}{\tau}}_{\substack{\text{spontaneous} \\ \text{emission into} \\ \text{the mode}}} .$$

$$(4.25b)$$

Here, A is a proportionality constant, d is the width
of the recombination region, and γ_s is the probability that an
emitted photon goes into the cavity mode. The photon lifetime
is given by

$$\frac{1}{\tau_{ph}} = \frac{c}{n} \left[\tilde{\alpha} + \frac{1}{2L} \ln\left(\frac{1}{R_1 R_2}\right) \right] , \qquad (4.26)$$

where c is the speed of light in vacuum, n is the refractive index at the lasing wavelength, and $\tilde{\alpha}$ is the effective internal absorption coefficient.

For a constant current density J, Eqs. (4.25a) and (4.25b) have a steady-state solution. Assume that at t = 0, the electron and photon populations in the active region deviate slightly from their equilibrium values by $(\Delta N_e)_0$ and $(\Delta N_{ph})_0$, respectively. Then, these transient electron and photon population deviations decay as follows:

$$\Delta N_e = (\Delta N_e)_0 \exp[-(a - i\omega_c)]t \qquad (4.27a)$$

$$\Delta N_{ph} = (\Delta N_{ph})_0 \exp[-(a - i\omega_c)]t \quad , \qquad (4.27b)$$

where a and the frequency ω_c are given by

$$a \simeq \frac{1}{2\tau}\left(\frac{J}{J_{th}} + 1\right) \qquad (4.28)$$

$$\omega_c = 2\pi f_{cr} \simeq \left[\frac{1}{\tau\tau_{ph}}\left(\frac{J}{J_{th}} - 1\right)\right]^{1/2} \quad . \qquad (4.29)$$

This model predicts that the coupled carrier-photon system will respond to any driving perturbation in a "resonant" way, with the response peaking at increasing frequency with increasing J/J_{th} (4.67). Both the damping term a and the resonance frequency f_{cr} increase with current.

While the model is crude and we have further simplified matters by linearizing the equations, it gives a semiquantitative description of many phenomena observed in laser diodes. Sudden laser turn-on, such as associated with pulsed operation, can result in oscillatory "ringing" which lasts a time $\approx a^{-1}$, Figure 4.26(a).

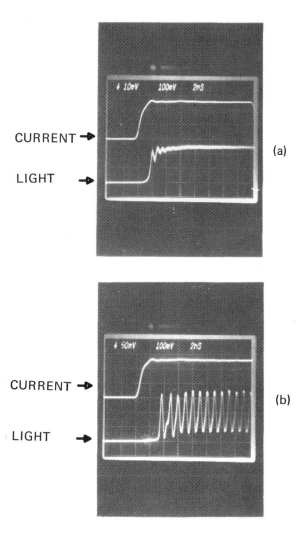

Figure 4.26. (a) Damped oscillatory behavior as measured in a DH
              AlGaAs laser diode as it is turned on. The delay
              in laser turn-on after application of the current
              is evident. (b) undamped oscillations which some-
              times appear in lasers that have degraded.

If a broadband analog modulation signal is applied to a laser, the response is frequency-dependent, as shown in Figure 4.27 (4.68). The laser response to intrinsic quantum noise fluctuations shows similar "resonance" effects. That these effects can occur in an important part of the frequency spectrum is shown by inserting typical values into Eqs. (4.28) and (4.29). For example, with $J = 1.2\ J_{th}$, $\tau = 2 \times 10^{-9}$ sec and $\tau_{ph} = 10^{-12}$ sec, $f_{cr} \approx 1.6$ GHz and $a^{-1} \approx 1.8$ nsec.

Not all experimental observations on such oscillatory effects can be explained by this simple model. A detailed treatment (4.69) shows that, at very high photon densities, the fluctuation effects are suppressed. Thus, the discussion above should not be taken to imply that high-frequency modulation of lasers is impossible. Indeed, AlGaAs DH laser pulse code modulation at 2.3 Gb/sec has been reported (4.66). Similar results are expected with InGaAsP/InP lasers (4.70).

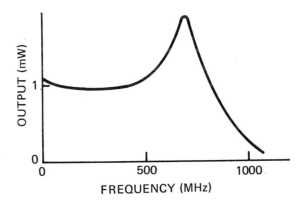

Figure 4.27. Experimental frequency response of an oxide-isolated AlGaAs DH laser with a sinusoidal modulation of 12 mA peak-to-peak and a 205 mA bias current (Ref. 4.68).

The theory given above predicts that all disturbances eventually damp out. However, it is possible to observe sustained pulsing from lasers as shown in Figure 4.26(b). One way that such undamped pulse trains can arise is from the phenomenon of *saturable absorption*. If a region in the laser cavity contains material whose absorption coefficient varies with the incident photon flux in a way that the absorption can *saturate*, an undamped instability can exist (4.71). Such absorption can arise from various defects in the laser material. Self-induced pulsations in diode lasers have been attributed to this effect (4.72). Anomalous pulsations after substantial laser degradation (4.73) may also be a result of saturable absorption associated with defects (4.74).

PROBLEMS FOR CHAPTER 4

1. Calculate the internal quantum efficiency at 300 K of a GaAs homojunction LED which contains 10^{16} cm^{-3} nonradiative recombination centers with a capture cross section $\sigma_t = 10^{-15}$ cm^2. The p-type recombination region has a doping level of 10^{17} cm^{-3}. Calculate for: (a) low injection level, (b) very high injection level. What would be the efficiency of a Si LED under the same conditions?

2. Derive an expression for the density of injected carrier pairs as a function of current density in a double heterojunction LED containing nonradiative recombination centers.

3. How would Equation 4.7 be modified if we analyze a LOC structure instead of a simple double heterojunction laser?
    Hint: The mode guiding region in the LOC laser is designed to be wider than the recombination region, thus reducing the fraction of the radiation in the recombination region.

4. Calculate the internal quantum efficiency in a double heterojunction LED where $\Delta a_o / a_o = 0.5\%$. Assume GaAs in the recombination region with a p-type doping level of 10^{18} cm^{-3}.

5. Derive an expression for the power conversion efficiency of a laser diode as a function of current and calculate values if $I_{th} = 100$ mA and the series resistance is 1 ohm. What is the efficiency as a function temperature if the thermal resistance is 20 K/W? At what current does the power conversion efficiency peak?

6.   Calculate the threshold current between 0°C and 100°C of
        a laser diode where T_o = 100 K.   The electrical series
        resistance is 1 Ω and the thermal resistance is 20 K/W.
        At T = 0°C, the threshold current is 100 mA.

7.   Calculate the highest number lateral mode capable of
        propagating in a GaAs laser diode where W = 10 μm and
        the lateral walls are etched (i.e., GaAs/air interface).

8.   Calculate the bandwidth f_c for a double heterojunction
        LED with GaAs in the recombination region ($p_o = 10^{18}$
        cm^{-3}) at low injection level if $\Delta a_o/a_o$ = 0.1%.

9.   Compare the theoretical f_c of a GaAs and a Si LED with
        other parameters constant.

10.   What is the maximum practical on-off modulation frequency
        of a laser diode where τ = 1 ns.   (a) With no current
        bias.   (b) With I_b = 1/2 I_{th}.

REFERENCES

4.1   H. Kressel and H. Nelson, *RCA Review 30*, 106 (1969).

4.2   I. Hayashi, M.B. Panish, and P.W. Foy, *IEEE J. Quantum Electron. QE-5*, 211 (1969).

4.3   Zh. I. Alferov, V.M. Andreev, E.L. Portnoi, and M.K. Trukan, *Sov. Phys. Semiconductors 3*, 1107 (1970).

4.4   (a)   M.H. Pilkuhn, *Phys. Status Solidi 25*, 9 (1969);

       (b)   A. Yariv, *Quantum Electronics* (John Wiley and Sons, Inc., New York, 1967);

       (c)   *GaAs Lasers*, C.H. Gooch, Ed. (John Wiley and Sons, Inc., New York, 1969);

       (d)   H. Kressel, a chapter in *Advances in Lasers, Vol. 3*, A.K. Levine and A.J. DeMaria, Eds. (Marcel Dekker, New York, 1971);

       (e)   J.I. Pankove, *Optical Processes in Semiconductors* Prentice-Hall, New York, 1971);

       (f)   H. Kressel and J.K. Butler, *Semiconductor Lasers and Heterojunction LEDs* (Academic Press, New York, 1977).

       (g)   P.G. Eliseev, *Sov. J. Quantum Electron. 2*, 505 (1973);

       (h)   H.C. Casey, Jr. and M.B. Panish, *Heterostructure Lasers* (Academic Press, New York, 1978).

4.5   Y.P. Varshni, *Phys. Status Solidi 19*, 353 (1964).

4.6   H. Kressel, H.F. Lockwood, F.H. Nicoll, and M. Ettenberg, *IEEE J. Quantum Electron. QE-9*, 383 (1973).

4.7   H. Kressel, J.K. Butler, F.Z. Hawrylo, H.F. Lockwood, and M. Ettenberg, *RCA Review 32*, 393 (1971).

4.8   H.F. Lockwood, H. Kressel, H.S. Sommers, Jr., and F.Z. Hawrylo, *Appl. Phys. Lett. 17*, 499 (1970).

4.9   G.H.B. Thompson and P.A. Kirkby, *IEEE J. Quantum Electron. QE-9*, 311 (1973).

4.10   J.J. Tietjen, *Ann. Rev. Mate. Sci. 3*, 317 (1973).
       Published by Annual Reviews, Inc., Palo Alto, Calif.

4.11   A comprehensive review is presented by H. Kressel and
       H. Nelson, "Properties and Applications of III-V Compound
       Films Deposited by Liquid Phase Epitaxy," in *Physics of
       Thin Films*, G. Hass, M. Francombe, and R.W. Hoffman, Eds.,
       Vol. 7 (Academic Press, New York, 1973).

4.12   D.B. Holt, *J. Phys. Chem. Solids 27*, 1053 (1966);
       H. Kressel, *J. Electron. Materials 4*, 1081 (1975).

4.13   M. Ettenberg and G.H. Olsen, *J. Appl. Phys. 48*, 4275 (1977).

4.14   R.D. Burnham, P.D. Dapkus, N. Holonyak, Jr., D.L. Keune,
       and H.R. Zwicker, *Solid-State Electron. 13*, 199 (1970).

4.15   M. Ettenberg and H. Kressel, *J. Appl. Phys. 47*, 1538 (1976).

4.16   H. Kressel in *Semiconductors and Semimetals*,
       R.K. Willardson and A.C. Beer, Eds., Vol. 14 (Academic
       Press, New York, 1979).

4.17   I. Hayashi, M.B. Panish, P.W. Foy, and S. Sumski, *Appl.
       Phys. Lett. 17*, 109 (1970).

4.18   H. Kressel and F.Z. Hawrylo, *Appl. Phys. Lett. 17*, 169
       (1970).

4.19   J.C. Dyment, *Appl. Phys. Lett. 10*, 84 (1967).

4.20   H. Yonezu, I. Sakuma, K. Kobayashi, T. Kamejima, M. Ueno,
       and Y. Nannichi, *Japan. J. Appl. Phys. 12*, 1585 (1973).

4.21   J.C. Dyment, L.A. D'Asaro, J.C. North, B.I. Miller, and
       J. E. Ripper, *Proc. IEEE (Letters) 60*, 726 (1972).

4.22   B.W. Hakki, *J. Appl. Phys. 44*, 5021 (1973).

4.23   I. Ladany and H. Kressel, *Appl. Phys. Lett. 25*, 708 (1974).

4.24   H. Kressel and I. Ladany, *RCA Review 36*, 230 (1975).

4.25   R.W. Keyes, *IBM J. Res. Dev. 15*, 401 (1971).

4.26   T. Tsukada, *J. Appl. Phys. 45*, 4899 (1974).

4.27   B.L. Frescura, C.J. Hwang, H. Luechinger, and J.E. Ripper,
       Appl. Phys. Lett. 31, 770 (1977).

4.28   D.R. Scifres, W. Streifer, and R.D. Burnham, *Appl. Phys. Lett. 32*, 231 (1978).

4.29   N. Matsumoto and H. Kawaguchi, *Japan. J. Appl. Phys. 16*, 1885 (1977).

4.30   P.J. Dewaard, *Electron. Lett. 13*, 400 (1977).

4.31   N. Chinone, K. Saito, R. Ito, K. Aiki, and N. Shige, *Appl. Phys. Lett. 35*, 513 (1979).

4.32   W. Tsang and R.A. Logan, *IEEE J. Quantum Electron. QE-15*, 451 (1979).

4.33   K. Aiki, N. Nakamura, T. Kuroda, J. Umeda, R. Ito, N. Chinone, and M. Maeda, *IEEE J. Quantum Electron. QE-14*, 89 (1978).

4.34   H. Kumabe, T. Tanaka, H. Namizaki, M. Ishii, and W. Susaki, *Appl. Phys. Lett. 33*, 38 (1978).

4.35   P.A. Kirkby and G.H.B. Thompson, *J. Appl. Phys. 47*, 4578 (1976).

4.36. R.D. Burnham, D.R. Scifres, W. Streifer, and S. Peled, *Appl. Phys. Lett. 35*, 734 (1979).

4.37   D. Botez and D. Zorey, *Appl. Phys. Lett. 32*, 761 (1978).

4.38   D. Botez, *Appl. Phys. Lett. 33*, 872 (1978).

4.39   C.A. Burrus and B.I. Miller, *Opt. Commun. 4*, 307 (1971).

4.40   J.P. Wittke, M. Ettenberg, and H. Kressel, *RCA Review 37*, 159 (1976).

4.41   H. Kressel and M. Ettenberg, *Proc. IEEE 63*, 1360 (1975).

4.42   M. Ettenberg, H. Kressel, and J.P. Wittke, *IEEE J. Quantum Electron. QE-12*, 360 (1976).

4.43   J. Conti and M.J.O. Strutt, *IEEE J. Quantum Electron. QE-8*, 815 (1972).

4.44   T.P. Lee and C.A. Burrus, *IEEE J. Quantum Electron. QE-8*, 370 (1972).

4.45   Y.S. Liu and D.A. Smith, *Proc. IEEE 63*, 542 (1975). J. Wittke independently derived this expression for

heterojunction structures. We are indebted to
Dr. Wittke for the $P(\omega)$ measurements given here.

4.46   H. Kressel and M. Ettenberg, *Appl. Phys. Lett. 23*, 511
        (1973).

4.47   J.J. Hsieh, *Appl. Phys. Lett. 28*, 283 (1976).

4.48   G.H. Olsen, C.J. Nuese, and M. Ettenberg, *Appl. Phys.
        Lett. 34*, 262 (1979).

4.49   S. Akiba, Y. Itaya, K. Sakai, T. Yamamoto, and
        Y. Suematsu, *Trans. IEEE Japan E-61*, 124 (1978).

4.50   Extensive data are given by P.G. Eliseev in *Semiconductor
        Light Emitters and Detectors*, A. Frova, Ed. (North-
        Holland Publishing Co., Amsterdam, 1973).

4.51   M. Ettenberg, H.S. Sommers, Jr., H. Kressel, and
        H.F. Lockwood, *Appl. Phys. Lett. 18*, 571 (1971).

4.52   B.W. Hakki and R. Nash, *J. Appl. Phys. 45*, 3907 (1974).

4.53   T. Yuasa, M. Ogawa, K. Endo, and H. Yonezu, *Appl. Phys.
        Lett. 32*, 119 (1978).

4.54   I. Ladany, M. Ettenberg, H.F. Lockwood, and H. Kressel,
        Appl. Phys. Lett. 30, 87 (1977).

4.55   A comprehensive discussion of the literature until 1973
        was presented by H. Kressel and H.F. Lockwood, *J. de
        Physique, C3, Suppl. 35*, 223 (1974).

4.56   H. Kressel and N.E. Byer, *Proc. IEEE 38*, 25 (1969).

4.57   W.D. Johnston and B.I. Miller, *Appl. Phys. Lett. 23*, 192
        (1973).

4.58   M. Ettenberg, H. Kressel, and H.F. Lockwood, *Appl. Phys.
        Lett. 25*, 82 (1974).

4.59   B.C. DeLoach, B.W. Hakki, R.L. Hartman, and L.A. D'Asaro,
        Proc. IEEE 61, 1042 (1973).

4.60   P. Petroff and R.L. Hartman, *Appl. Phys. Lett. 23*, 469
        (1973).

4.61   R.D. Gold and L.R. Weisberg, *Solid-State Electron. 7*, 811
        (1964).

4.62   M. Ettenberg and H. Kressel, *Appl. Phys. Lett.* *26*, 478 (1975).

4.63   R.L. Hartman and R.W. Dixon, *Appl. Phys. Lett.* *26*, 239 (1975).

4.64   M. Ettenberg and H. Kressel, *IEEE J. Quantum Electron.* *QE-16*, 170 (1980).

4.65   For a review, see G. Arnold and P. Russer, *Appl. Phys.* *14*, 255 (1977).

4.66   G. Arnold, P. Russer, and K. Petermann, "Modulation of Laser Diodes," in *Semiconductor Devices for Optical Communication*, H. Kressel, Ed. (Springer-Verlag, New York, 1980).

4.67   H. Haug, *Phys. Rev.* *184*, 338 (1969).

4.68   A.R. Goodwin, P.R. Selway, M. Pion, and W.O. Bourne, *AGARD Conf. Proc.* *219* (1977).

4.69   D.E. McCumber, *Phys. Rev.* *141*, 306 (1966).

4.70   S. Akiba, K. Sakai, and T. Yamamoto, *Electron. Lett.* *14*, 197 (1978).

4.71   See, for example, T.P. Lee and C.H.R. Roldan, *IEEE J. Quantum Electron.* *QE-6*, 339 (1970).

4.72   T. Ohmi, T. Suzuki, and M. Nishimaki, *Oyo Buturi 42*, 102 (1972) (supplement).

4.73   D.J. Channin, M. Ettenberg, and H. Kressel, *J. Appl. Phys.* *50*, 6700 (1979).

4.74   E.S. Yang, P.G. McMullin, A.W. Smith, J. Blum, and K.K. Shih, *Appl. Phys. Lett. 24*, 324 (1974).

CHAPTER 5

PHOTODETECTORS FOR FIBER SYSTEMS

S.D. Personick

TRW Technology Research Center
El Segundo, California 90245

5.1   INTRODUCTION

Photodetectors are devices that convert optical power
to electric current. Since we are primarily interested in devices
that can be used in receivers for fiber-optic communication links,
contents of this chapter will be limited to photodiodes, which are
currently used in those applications. We can treat the subject
from several perspectives. First, we shall develop a physical
model for photodetection, including discussion of the considerations that have led to structures presently used for actual devices.
Next, we shall develop circuit models for various detector types,
so that we can understand how to incorporate these devices into
actual receiver circuits. Finally, we shall develop a statistical, or noise model, of the detection process in order to have
the necessary tools for predicting the performance of optical
receivers.

This chapter will provide the necessary background for
the next chapter on receiver design.

5.2   PHOTODETECTION

        In this section, we discuss photodetection beginning with
the familiar photoelectric effect (5.1-5.5). Consider the simple
vacuum photodiode shown in Figure 5.1. The photocathode will emit
electrons when illuminated by light having a sufficiently short
wavelength. The photon energy of the incident light must be large
enough to liberate electrons from the photocathode surface. Photons absorbed too deep below the surface will produce electrons
that tend to lose too much energy before they can reach the surface and escape (overcome the surface potential barrier). The
fraction of incident photons that produce liberated electrons
is called the detector quantum efficiency. By definition, the
maximum quantum efficiency is unity. Another equivalent measure
of the efficiency with which incident light is converted to current is called the responsivity. This is defined as the photocurrent, in amperes, divided by the incident light power, in
watts. Since the photocurrent is equal to the number of electrons emitted per second multiplied by the electron charge
$(1.6 \times 10^{-19}$ C), and since the incident light power is equal to
the number of photons per second incident multiplied by the
energy in a photon (about 2×10^{-19} J at 1 μm wavelength), it follows that the responsivity is equal to the quantum efficiency,
η, multiplied by e/hν (where e is the electron charge and hν is
the photon energy). Because of the way constants were histori-
cally defined, at 1 μm wavelength the responsivity is numeri-
cally about 0.8 times the quantum efficiency (in amperes/watt).
Curves for quantum efficiency and responsivity of some typical
photoemissive materials are shown in Figure 5.2.

        In addition to quantum efficiency or responsivity, the
speed of response of a photodetector is an important measure of
its performance. In the vacuum photodiode, the speed of
response is determined by the duration of the displacement

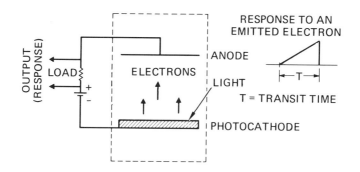

Figure 5.1.   Vacuum photodiode.

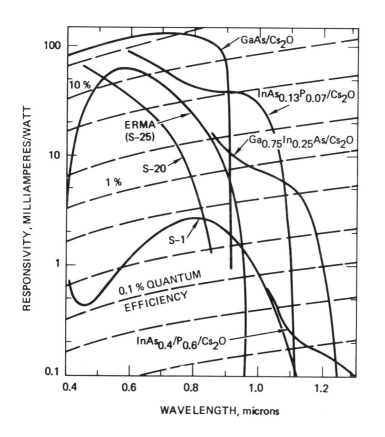

Figure 5.2.   Responsivity and quantum efficiency of typical
              photocathodes.

current that flows when an electron is liberated from the detector
surface (assuming that the delay for electrons to reach the surface once they are produced is relatively small). At any point during the flight of the electron from the cathode to the anode, the
displacement current produced is proportional to the component
of the velocity of the electron in the direction of the electric
field, multiplied by the electric field strength at the location
of the electron. For a vacuum photodiode with a uniform electric
field between the cathode and the anode, the displacement current
increases as the electron accelerates.

Vacuum photodiodes illustrate the basic physics of photodetection and the important parameters of quantum efficiency and
response speed. Typically, they are too bulky, expensive, and
difficult to use in communications applications (for fiber optic
systems) except occasionally in system test instruments.

5.3   PN AND PIN PHOTODIODES

The simplest solid-state photodiode is illustrated in
Figure 5.3. It consists of a backbiased PN junction in series
with a load resistor. The reverse voltage causes mobile holes
and electrons to move away from the junction, leaving behind
immobile positive donor ions and negative acceptor ions. These
immobile charges produce an electric field in the vicinity of
the junction as shown. The region in which the field exists is
called the depletion region. As shown in Figure 5.4, when the
detector is illuminated by incident light, hole-electron pairs
will be produced in a region defined as the absorption region.
The depth of the absorption region (light penetration) depends
on how strongly light is absorbed at the given wavelength in the
particular material used to make the diode.

Hole-electron pairs moving within the device produce
a displacement current proportional to the product of the component of carrier velocity in the direction of the local field and
the local field strength.

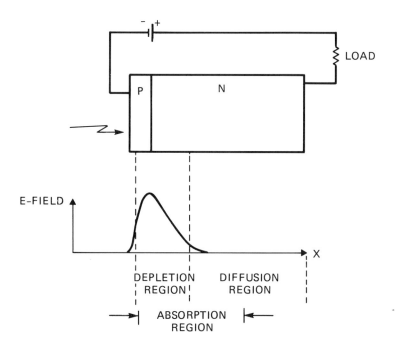

Figure 5.3.   PN diode.

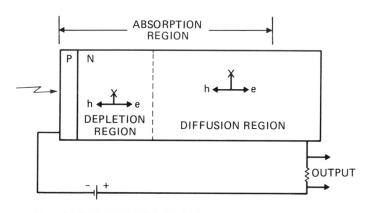

X = HOLE-ELECTRON PAIR CREATED

Figure 5.4.   Carrier generation.

In the depletion region, carrier velocities are high
(saturation limited) as are the field strengths. In the diffusion
region, carriers move randomly with relatively low velocities, and
in the presence of essentially no field. Thus, most of the displacement current produced by a hole-electron pair occurs as a
short duration pulse when the carrier or carriers are moving
through the depletion region. Carriers produced within the depletion region produce an immediate response. Carriers produced
within the diffusion region produce a randomly delayed response
(or a "diffusion tail"). To make a high speed device, it is necessary for the depletion region to enclose the absorption region.
This is accomplished by increasing the bias voltage or by decreasing the doping in the N material. Since practical constraints
limit the applied bias voltage, doping levels are reduced until
the N region is almost intrinsic (I type). For the practical purpose of making an ohmic contact, a highly doped N region must be
added as shown in Figure 5.5. This results in the familiar PIN
structure. Figure 5.6 shows a practical silicon PIN detector
geometry suitable for fabrication by diffusion techniques. It
should be noted that light absorbed in the highly doped N region
need not produce a diffusion tail since, in the highly doped material, hole-electron pairs recombine before they can produce a displacement current.

The PIN detector geometry is a tradeoff between quantum
efficiency and speed of response. If the I region is very wide,
then the transit time will be long, and, therefore, the response
speed will be slow. If the I region is too narrow, then much
of the light may be absorbed beyond the I region, where no useful photocurrent is produced. For wavelengths below 0.9 µm, where
silicon is highly absorbent, diodes can be fabricated with less
than 0.5 nsec response speeds and essentially 100% quantum efficiency (limited by reflections from the diode surface). For
longer wavelengths, the silicon becomes transparent, and the

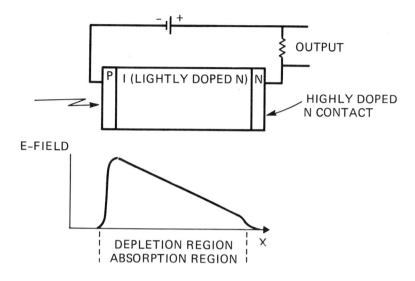

Figure 5.5.   Pin diode.

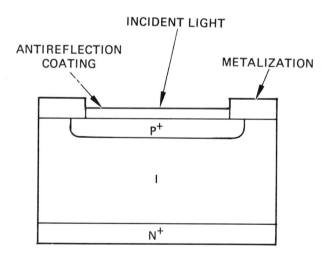

Figure 5.6.   Typical pin diode geometry.

quantum efficiency drops off rapidly. Other materials such as germanium and quaternary compounds (InGaAsP) are being investigated for longer wavelength devices. Since the other materials are not as well understood as silicon, problems with uniformity, leakage, etc., need to be resolved. Figure 5.7 shows the quantum efficiency and responsivity of some typical high-speed photodiodes versus incident light wavelength.

5.4   AVALANCHE PHOTODIODES

If one considers an ideal PIN photodiode in which every incident photon produces a hole-electron pair, then at 1 μm wavelength, the responsivity is, at most, about 0.8 A/W. As indicated in Chapter 6, typical high sensitivity digital receivers operate with input power levels as low as a few nanowatts. Thus with a PIN detector, the photocurrents produced would be a few nanoamperes or less. Such small photocurrents would be severely corrupted by amplifier noise in typical receivers. Therefore, it is desirable to devise a means for generating more than one electron-hole pair for each detected photon. An avalanche photodiode is shown schematically in Figure 5.8. (Suggested references discussing these devices are 5.4-5.10.)

The structure of an avalanche photodiode (5.4,5.5) includes a high field multiplication region (adjacent to a lightly doped depletion region) where the doping levels are high. When the device is reverse biased a field profile as shown in Figure 5.8 is produced. In the high-field region, carriers move at sufficient average velocities so that occasionally ionizing collisions occur. The new carriers produced in these ionizing collisions, in turn, can produce additional carriers by the same mechanism. As long as the process is stable (self-terminating) a single hole-electron pair generated by photon absorption can produce tens or hundreds of additional hole-electron pairs, thereby increasing the effective responsivity of the device.

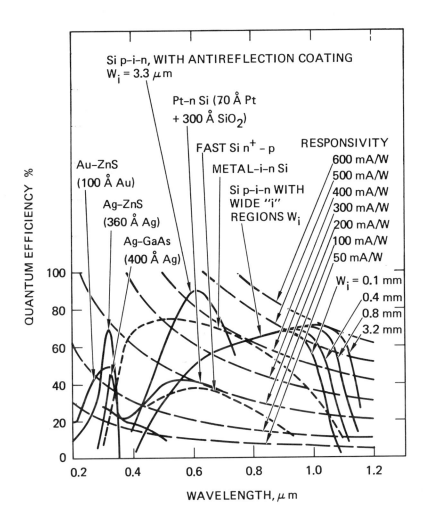

Figure 5.7. Quantum efficiency of typical high-speed photodiodes.

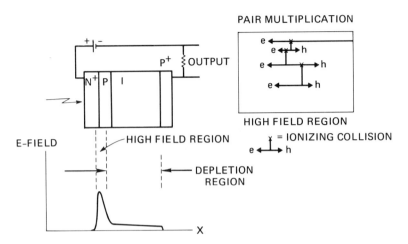

Figure 5.8.   Avalanche photodiode.

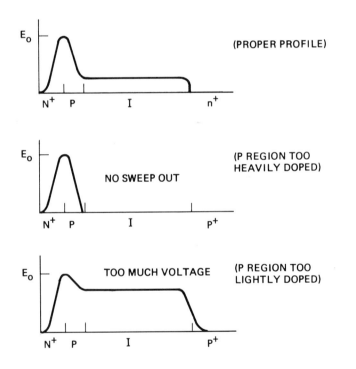

Figure 5.9.   Doping versus field profile in an APD.

The avalanche multiplication mechanism is not ideal, because the exact number of hole-electron pairs produced by a single absorbed photon is not predictable (5.6-5.10). This randomness in the multiplication process produces an effective noise which limits the sensitivity of receivers incorporating APDs. However, there are ranges of multiplications (gains) where in the presence of amplifier noise, the APD offers significant benefits relative to the PIN detector (as discussed in Chapter 6). To minimize the randomness or unpredictability of the avalanche gain, the material and device geometry must be as uniform as possible. In addition, it can be shown (5.6,5.9) that the best statistics (lowest effective noise) result when the more strongly ionizing carrier enters the multiplication (high field) region from the depletion region. Thus we would like most of the light absorption to take place in the depletion region. Beyond this, the ratio of ionization probability per unit length of motion in the high-field region for holes and electrons is an important parameter. The smaller this ratio (weaker-to-stronger ionizing carrier) the better.

The geometry and doping levels of the device must be chosen carefully to produce a fast device that operates at a reasonable reverse-bias voltage. (Refer to Figure 5.9.)

If the reverse bias is increased from zero, at first the multiplication region will begin to become depleted of mobile carriers. This will cause the field to increase in this region. As the bias is increased, eventually the multiplication region will be completely depleted and the I region will begin to become depleted. If the doping is too heavy in the multiplication region, the E field may grow large enough to cause infinite multiplication before the I region begins to deplete. This will result in a slow device (since the voltage must be maintained below the level which produces instability). On the other hand, if the doping is too low in the multiplication region, the I

region will deplete too soon; therefore, the applied voltage
required to initiate multiplication will be too high.

Assuming that the I region is depleted, and not too long,
the response speed of an APD is limited by the duration of the
multiplication process. Typical APDs exhibit a gain-bandwidth
product where higher multiplications take longer to transpire.
Fortunately, silicon APDs are available with gain capabilities
in excess of 100 and response speeds below 0.5 nsec.

Figure 5.10 illustrates a typical silicon APD geometry.
The guard ring prevents premature avalanche breakdown at the
device edges by keeping the field strengths at or below those
in the interior of the device. Again, for best noise performance, extreme uniformity and a low ratio of ionization probabilities per unit length of the carriers (in the high-field region)
is essential.

As in PIN detectors, silicon APDs lose their quantum
efficiency as wavelengths exceed 0.9 μm. Producing APDs from
other materials with low leakage, high uniformity, and good multiplication statistics continues to be a formidable challenge.

5.5   PIN PHOTODIODE CIRCUIT MODEL

In a typical application, a PIN photodiode would be interconnected to the amplifier input (load) as shown in Figure 5.11.
The power supply, bypassed by an appropriate large capacitor,
is placed in series with the diode and a load circuit. The resistor R_L provides a dc path back to the battery. The amplifier
(load) typically can be modeled by a resistor and capacitor in
parallel. The reverse biases for silicon detectors vary from 10
to >100 V, depending on the device response speed and the length
of the I region. Under typical operating conditions, most of the
reverse bias appears across the detector, even if the dc return

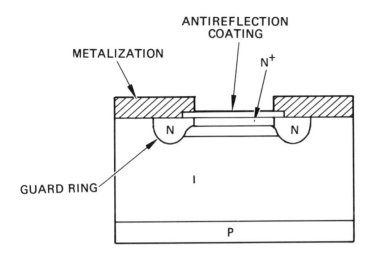

Figure 5.10. Typical APD geometry.

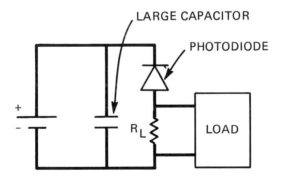

Figure 5.11. Photodiode circuit.

resistance is very large. The responsivity is linear over many decades of optical signal levels, although the response speed may decrease if high currents cause a significant drop in voltage across the load, thus decreasing the field levels in the detector. The limiting factor on allowable signal levels is typically device heating or the ability of the power supply to deliver the necessary current.

As carriers are generated and separate in the device, they create a voltage across the diode junction capacitance, which is discharged through the load circuitry. An ac equivalent circuit for fiber applications is shown in Figure 5.12. We see that the diode is essentially a current source in parallel with its junction capacitance (typically a few picofarads). A small series resistance is shown (a few ohms) that is negligible in most applications. There is also some shunt resistance, which typically is large enough to be neglected except at very low frequencies. (The load circuitry, i.e., the amplifier to be used with the photodiode, is discussed in Chapter 6.)

The optimal design of an amplifier is based on a number of practical considerations, as well as the major consideration of amplifier noise. Assume the following suboptimal (noise) amplifier shown in Figure 5.12. It is modeled as a resistor in parallel with a capacitor followed by an ideal, infinite impedance amplifier. If the amplifier has gain $A(f)$, where f is a baseband frequency, then the output voltage at frequency f in response to a component in the detector current at frequency f is given by the simple formula

$$V(f) = I(f)Z(f)A(f) \quad , \tag{5.1}$$

where $Z(f)$ is the impedance of the load at frequency f,

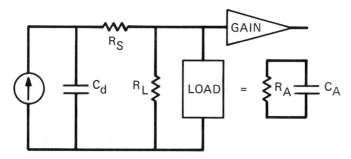

C_d = DIODE JUNCTION CAPACITANCE

R_S = DIODE SERIES RESISTANCE

R_L = PHYSICAL LOAD RESISTOR

R_A = AMPLIFIER INPUT RESISTANCE

C_A = AMPLIFIER SHUNT CAPACITANCE

C_T = $C_A + C_d$

R_T = $1/\left(\dfrac{1}{R_L} + \dfrac{1}{R_A}\right)$

Figure 5.12.   Equivalent circuit.

$$Z(f) = \frac{1}{1/R_T + j2\pi f C_{Total}} \qquad . \qquad (5.2)$$

Furthermore, if the optical power falling on the detector
is p(t) watts and if the detector responsivity is R (amperes/watt),
then the amplifier output voltage is

$$v(t) = p(t)R * h_{diode}(t) * h_{amplifier-load}(t) , \qquad (5.3)$$

where

        * indicates the convolution of two impulse responses

        $h_{diode}(t)$ is the detector impulse response (Fourier
        transform of the detector frequency response)

        $h_{amplifier-load}(t)$ is the load-amplifier impulse
        response (Fourier transform of Z(f)A(f)).

It is important to recognize that p(t), the received
optical power, varies in time at a baseband rate (according to
the modulation) not at an optical rate. If p(t) varies slowly,
then v(t) is just a scaled replica of p(t).

For example, consider Figure 5.13. We assume that p(t)
varies sinusoidally about some mean value with peak amplitude
(of the sinusoidal part) of 1 μW, and at the frequency 1 MHz.
We assume a detector responsivity of 1 A/W. We further assume
that the detector frequency response is unity to frequencies well
beyond 1 MHz (no detector rolloff). The amplifier input impedance is dominated by the 50 Ω resistor at these frequencies. The
amplifier gain is 20 dB. Thus, the amplifier output voltage is
a sinusoid with a peak value of 500 μV. Clearly, further amplification is needed to process this signal.

The circuit model described above assumes that the detector output current is a linearly filtered version of the received

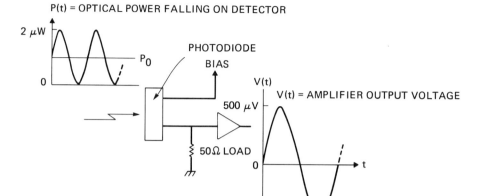

Figure 5.13.   Example: sinusoidal modulation.

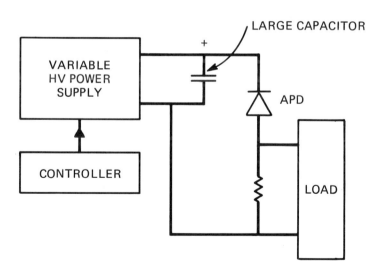

Figure 5.14.   Avalanche photodiode circuit.

optical power. For PIN detectors, this is an excellent
approximation, violated only at optical power levels that are
too high for consideration in fiber communication applications.

5.6   AVALANCHE PHOTODIODE CIRCUIT MODEL

         Much of the preceding discussion of the PIN diode cir-
cuit model applies to avalanche diodes. Figure 5.14 shows an
avalanche diode biasing circuit in schematic form. The important
distinction between this circuit and the PIN circuit is the very
strong dependence of the device responsivity on the bias voltage.
A typical variation of avalanche detector responsivity with bias
voltage for various temperatures is shown in Figure 5.15. That is
the reason for the temperature-compensating power supply. As the
temperature varies, the detector breakdown voltage varies, and,
therefore, the proper bias required to maintain a fixed avalanche
gain varies. This proper bias may be hundreds of volts or more for
typical avalanche detectors. The design of the voltage control circuit depends on the application. In some cases, the average current flowing in the bias circuit can be monitored and fed back to
the controller. Sometimes, an auxiliary avalanche diode, matched
to the active diode, is monitored, to determine its avalanche gain.
If the matching is good, then the bias voltage, which is applied
to both diodes, can be varied to produce the proper gain in the
auxiliary diode and, therefore, in the active diode. In digital
repeater applications (see Chapter 6) the peak detected receiver
output can control the APD bias as part of the automatic gain
control loop. The need to provide a controlled high-voltage bias
is a serious drawback of avalanche photodiodes. However, this
is frequently offset by the enhanced responsivity resulting from
the carrier multiplication.

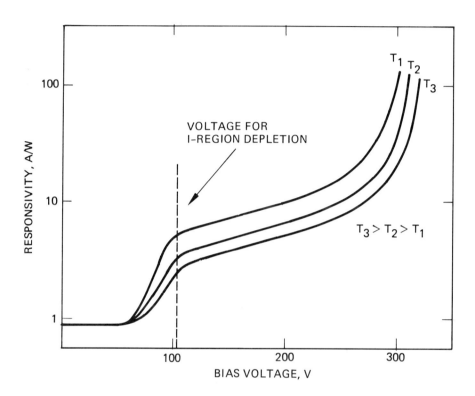

Figure 5.15. Responsivity versus voltage for a typical APD.

An equivalent circuit for an avalanche photodiode is given in Figure 5.16. It is identical with the PIN diode circuit with the following caution. Since the avalanche multiplication is field-level sensitive, nonlinearity (saturation) can occur if the applied optical signal variations are very large. Usually, avalanche detectors are used to achieve the ultimate in sensitivity, implying small optical signals. However, where dynamic range requirements are excessive, caution must be exercised.

5.7 NOISE IN PHOTODIODES (WITHOUT AVALANCHE GAIN)

In the previous sections, we discussed the concept of responsivity. When a given power p(t)(watts) was incident upon the photodiode, a certain average current i(t) was generated by the diode. This current was proportional to the power, with the responsivity (amperes/watt) being the proportionality constant.

We know that the diode current consists of the sum of the displacement currents of individual hole-electron pairs generated within the device (see Figure 5.17). The times at which these hole-electron pairs are generated are not precisely predictable. Thus, although the average current pulse response (defined by adding up a large number of current pulse responses and dividing by the number which we added) looks like the input power pulse, any individual pulse response differs from the average by some unpredictable amount. We call the difference between the actual pulse in any given response, and the average pulse response, a signal-dependent noise e(t). This noise is signal dependent because its statistics are a function of the average pulse. To design receivers that perform reliably, we must know something about the statistics of this signal-dependent noise. In the next section, we formulate a noise model for PIN and avalanche photodiodes.

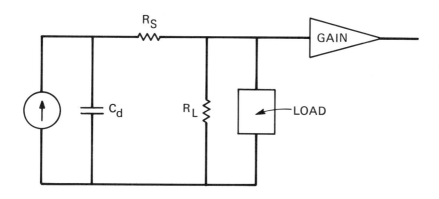

Figure 5.16.   APD equivalent circuit.

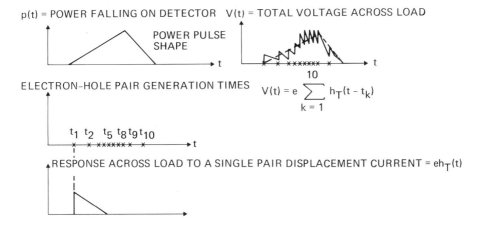

$p(t)$ = POWER FALLING ON DETECTOR $V(t)$ = TOTAL VOLTAGE ACROSS LOAD

POWER PULSE SHAPE

ELECTRON-HOLE PAIR GENERATION TIMES

$$V(t) = e \sum_{k=1}^{10} h_T(t - t_k)$$

t_1 t_2 t_5 t_8 t_9 t_{10}

RESPONSE ACROSS LOAD TO A SINGLE PAIR DISPLACEMENT CURRENT = $eh_T(t)$

Figure 5.17.   Sum of displacement currents of individual hole-
electron pairs generated within device.

5.8   MATHEMATICAL NOISE MODEL

We begin our model by assuming that a fixed pulse of
optical power p(t) watts is incident upon the photodiode.   In
response to this power pulse, hole-electron pairs are generated
in the diode at times $\{t_n\}$.

These hole-electron pairs produce displacement currents
which build up a voltage across the diode load. For simplicity,
we shall assume that the displacement current produced by a hole-
electron pair is the same as that produced by any other hole-
electron pair.

By referring to Figure 5.17, we see the voltage developed
across the load is given by the formula

$$V_{load}(t) = e \sum_{n=1}^{N} h_T(t - t_n) . \qquad (5.4)$$

In Eq. (5.4), $eh_T(t)$ is the overall impulse response of the diode
and the load to an electron-hole pair generated at time t = 0.
The total number of hole-electron pairs generated is N.   Both
N and the set of generation times $\{t_n\}$ are random quantities.
It has been shown that these random generation times form what
is referred to by statisticians as a Poisson random process, with
a time-varying rate (5.14). Basically, this means that if we divide
the time axis into small intervals of length dt, as shown in
Figure 5.18, then in any interval, either 1 or 0 hole-electron
pairs will be created within the diode.   The probability that one
pair is created, is given by $\lambda(t)dt$, where $\lambda(t) = (\eta/h\nu)p(t)$.
Clearly $\lambda(t)$ = average number of electrons generated per second.
The probability, that no pair is created, is given by $1 - \lambda(t)dt$.

These formulas assume that the intervals are so small that the
probability that more than one pair is created in an interval is
negligible. In addition, in the statistical model, it is assumed
that whether a given interval gives rise to a hole-electron pair
is independent of whether any other interval gives rise to a pair.

From these few assumptions, a number of important conclusions follow. First, the total number of hole-electron pairs
created during the interval of time $[t, t + T]$ is a random variable having the following probability distribution:

$$\text{Prob}(N = n) = \frac{\Lambda^n e^{-\Lambda}}{n!} , \qquad (5.5)$$

where

$$\Lambda = \int_t^{t+T} \lambda(t)\,dt = \int_t^{t+T} \frac{\eta}{h\nu}\, p(t)\,dt . \qquad (5.6)$$

This is the familiar Poisson distribution with mean value of N
equal to Λ.

Furthermore, referring to Eq. (5.4) the average voltage
across the load can be derived as (5.14)

$$V_{\text{load average}} = \frac{\eta e}{h\nu} \int h_T(t - \tau)p(\tau)\, d\tau . \qquad (5.7)$$

In addition, at any time t, the mean square deviation
of the voltage developed across the load from its average value
is (5.14)

$$\left\langle \left[V(t)_{\text{load}} - V_{\text{load average}}(t) \right]^2 \right\rangle = \frac{\eta}{h\nu}\, e^2 \int h_T^2(t - t)p(\tau)\,d\tau . $$
$$(5.8)$$

This mean-square deviation is a crude measure of the
randomness of the load voltage.

In Eq. (5.8) we have considered the randomness in the load voltage caused by the random photodiode hole-electron pair generation times. The amplifier output voltage depends on this voltage plus noise from the diode biasing resistor (Johnson noise) and from the amplifier internal noise sources. The amplifier output voltage can be written as

$$\left\langle V_{amp\ output}(t) \right\rangle = \frac{ne}{h\nu} \int h_S(t - \tau)p(\tau)d\tau \qquad (5.9)$$

$$\left\langle \left(V_{amp\ output} - \left\langle V_{amp\ output} \right\rangle \right)^2 \right\rangle = \left[\frac{ne^2}{h\nu} \int h_S^2(t - \tau)p(\tau)d\tau \right] + N_{thermal}^2 \quad , \qquad (5.10)$$

where $h_S(t)$ is the convolution of $h_T(t)$ with the amplifier impulse response, and $N_{thermal}^2$ is the mean-squared thermal noise at the amplifier output caused by the biasing resistor and the internal amplifier noise sources. It should be noted that this output thermal noise variance depends on the amplifier type being used and on the frequency response of any filters included in the amplifier. This noise is often erroneously assumed to be proportional to the amplifier bandwidth. Assume for now that the amplifier is fixed so that $N_{thermal}^2$ is fixed.

From Eqs. (5.5) and (5.9), we see that for a constant $p(t) = P_o$ the signal-to-noise ratio (SNR) defined as the square of the average amplifier output signal divided by the mean-square deviation from the average is given by

$$\frac{\left\langle V_{amp\ output} \right\rangle^2}{\left\langle (V_{amp\ output} - \left\langle V_{amp\ output} \right\rangle)^2 \right\rangle} = SNR$$

$$= \frac{\left(\frac{ne}{h\nu} P_o \int h_S(t)dt \right)^2}{\frac{ne^2}{h\nu} P_o \int h_S^2(t)dt + N_{thermal}^2} \quad . \qquad (5.11)$$

If we choose $h_s(t)$ to be the impulse response of an ideal low-pass filter with bandwidth B, we obtain

$$SNR = \frac{\left(\frac{\eta e P_o}{h\nu}\right)^2}{\frac{2\eta e^2 P_o}{h\nu} B + N_{thermal}^2} \quad ,$$ (5.12)

where $N_{thermal}^2$ depends on B, the detector load circuit, and the amplifier being used.

We see from Eq. (5.12) that for a given thermal noise, $N_{thermal}^2$, there is a value of P_o = optical input power that gives unity SNR (average signal = rms "noise" at the amplifier output). This is referred to as the noise-equivalent power (NEP). The lower the NEP of a given detector-amplifier combination, the less optical power is required to obtain the desired SNR. In receivers without avalanche gain, and at SNRs near unity, the thermal noise $N_{thermal}^2$ dominates the shot noise (or quantum noise) in the denominator of Eq. (5.12). Thus, the NEP varies with the receiver bandwidth as the square root of $N_{thermal}$ varies with the receiver bandwidth. Since $N_{thermal}$ is sometimes proportional to the receiver bandwidth, NEP is often expressed as watts per root Hertz. However, unless the amplifier and load resistance are specified, NEP is not a meaningful parameter.

5.9   NOISE IN PHOTODIODES (WITH AVALANCHE GAIN)

To obtain a bigger detector output signal to combat amplifier and load thermal noises, avalanche gain can be used. Referring to Figure 5.19, we see the voltage developed across the load by the displacement currents flowing in the detector is given by

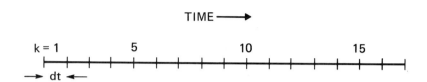

PROB "COUNT" IN INTERNAL dt $\cong \lambda$(kdt) dt

"NO COUNT" $\cong 1 - \lambda$(kdt) dt

λ(t) = INTENSITY OF THE PROCESS

Figure 5.18. Poisson random process.

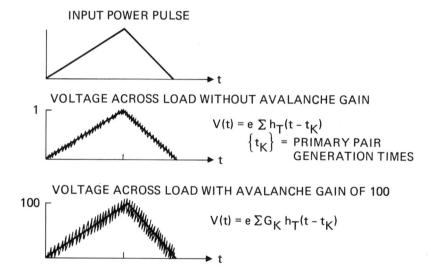

Figure 5.19. Effect of avalanche gain on detector statistics.

$$V_{load}(t) = e \sum_{n=1}^{N} G_n h_T(t - t_n) \quad . \tag{5.13}$$

In Eq. (5.13) t_n is the generation time of primary hole-electron pair n; G_n is the number of secondary pairs (including the primary) produced through the collision-ionization mechanism. For each n, G_n is a random variable with mean G. All the secondary pairs are assumed to be statistically independent. The probability distribution of the random gain G_n depends on the type of avalanche detector. In particular, it is a function of the ratio of hole collision ionization probability to electron collision ionization probability. That is, as holes and electrons drift through the high-field region of the detector, one carrier has a stronger probability per unit length of drift to produce a new hole-electron pair. The derivation of the detector statistics is too complicated to reproduce here. We shall only summarize a few important results.

It can be shown (5.6, 5.9) that for "good" multiplication statistics (minimum randomness) primary hole-electron pairs should be generated outside of the high-field region, so that the more strongly ionizing carrier drifts into the multiplication region.

If the ratio of ionization probabilities is k, then the mean square number of secondaries is related to the mean number of secondaries (mean gain) as follows (5.6, 5.9):

$$\langle G_n^2 \rangle = F(\langle G_n \rangle) \cdot \langle G_n \rangle^2 = F(G) \cdot G^2 \quad , \tag{5.14}$$

$$F(G) \cong kG + \left(2 - \frac{1}{G}\right)(1 - k) \quad . \tag{5.15}$$

For an ideal (deterministic) multiplication mechanism we would have F(G) = 1. Thus F is a measure of the degradation

caused by the randomness of the multiplication. Figure 5.20 shows
curves of F(G) versus G for various values of k. For good silicon detectors, k is between 0.02 and 0.03. From (5.14) and from
our previous results in Section 5.8, it can be shown that the
amplifier output voltage has the following statistical properties (5.14) (see Figure 5.19):

$$\left\langle V_{amp\ output}(t) \right\rangle = \frac{eG\eta}{h\nu} \int h_S(t - \tau)p(\tau)d\tau$$

$$\left\langle \left(V_{amp}(t)_{output} - \left\langle V_{amp}(t)_{output} \right\rangle \right)^2 \right\rangle = \frac{e^2 F(G)G^2\eta}{h\nu}$$

$$\int h_S^2(t - \tau)p(\tau)d\tau + N_{thermal}^2 \ . \tag{5.16}$$

Thus, for constant optical input power P_o, and an ideal lowpass
$h_S(t)$ with bandwidth B the SNR is given by the formula (which
is analogous to Eq. (5.12)):

$$SNR = \frac{\left\langle V_{amp\ output} \right\rangle^2}{\left\langle \left(V_{amp\ output} - \left\langle V_{amp\ output} \right\rangle \right)^2 \right\rangle}$$

$$= \frac{\left(\frac{\eta eGP_o}{h\nu} \right)^2}{\frac{2\eta e^2 G^2 F(G)P_o B}{h\nu} + N_{thermal}^2} \ . \tag{5.17}$$

We see that for fixed thermal noise $N_{thermal}^2$, if we set
the SNR to unity, there is some value of mean gain G which minimizes the required optical power P_o. That is, the NEP of the
avalanche detector-amplifier combination can be minimized by properly adjusting the mean avalanche gain. If the avalanche gain
is increased, then the average signal increases relative to the
thermal noise. However, if the avalanche gain is increased too
far, the first term in the denominator of Eq. (5.17) (randomly

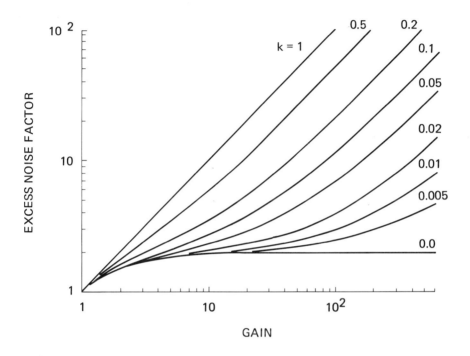

Figure 5.20. Excess noise factor versus gain F(G) versus G.

multiplied shot noise) begins to dominate. Further increase in
the average gain is detrimental because F increases. Figure 5.21
shows a typical plot of NEP versus G for a realistic amplifier
and detector, assuming a bandwidth requirement of 25 MHz. The
"quantum limit" curve corresponds to a situation where either
$N^2_{thermal} = 0$ or $F(G) = 1$ and $G \to \infty$. Note that the value of G,
which minimizes the NEP, is not necessarily the optimal value
of G to use at high SNRs.

Figure 5.22 shows the required optical power versus G
for various SNRs. Unity SNR is, of course, the NEP curve again.
We see that the higher SNRs require less gain. This follows from
Eq. (5.17).

Equations (5.14) through (5.17) give information about
"second moment" properties of the avalanche gain and the amplifier
output signal. For the design of digital communication system
receivers, more detailed statistical information is needed. This
is discussed further in Chapter 6 (see also Refs. 5.11-5.13).

5.10   DESIGNING A SIMPLE OPTICAL RECEIVER

Suppose we wish to build a receiver for analog-modulated
light at wavelength 0.9 μm. The received power waveform is

$$p(t) = P_o[1 + \gamma m(t)] \quad , \tag{5.18}$$

where P_o is the average received power, and m(t) is the message.
We assume m(t) has peak value unity and that the modulation index,
γ, is less than one. Furthermore, the message m(t) has bandwidth
B equal to 1 MHz. The receiver is shown in Figure 5.23. It consists of a silicon APD with a 50 Ω amplifier in series. This
amplifier provides the 50 Ω-dc resistance to complete the biasing
circuit. It should be pointed out that a 50 Ω amplifier is far
from optimal in this application for minimizing thermal noise.

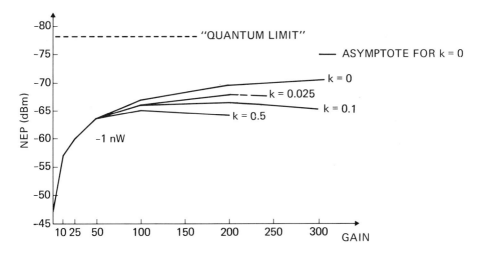

Figure 5.21. Noise equivalent power versus average gain typical
25 MHz receiver.

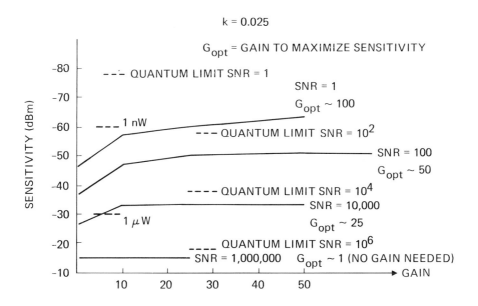

Figure 5.22. Sensitivity versus gain for typical 25 MHz receiver.

However, we use it in this example for simplicity. Assume that
the detector has a junction capacitance of 5 pF and that the amplifier has a shunt capacitance of 5 pF. The equivalent circuit
is shown in Figure 5.24. For this example, we characterize the
amplifier noise in terms of its noise figure, (3 dB at a 1 MHz
bandwidth). Thus the amplifier has internal noise sources that
produce as much output thermal noise as that of the Johnson noise
of the load resistance (50 Ω). We shall assume that the photodiode response is fast enough that the average diode output current waveform will follow the 1 MHz input power variations.

The total load impedance is 50 Ω for frequencies well
beyond 1 MHz (it rolls off at high frequencies). Furthermore,
the voltage produced across the load neglecting noise is

$$V_{load} = \frac{\eta eG}{h\nu} \; (50)[p(t)] \; . \tag{5.19}$$

We shall assume that the diode responsivity ($\eta eG/h\nu$)
at this wavelength is about $(0.5 \text{ A/W}) \times$ (avalanche gain).

From our results of Sections 5.8 and 5.9, we have the
amplifier output average voltage and average noise (the noise
given by Eq. (5.16) with p(t) set to its average value P_o) given
by

$$\left\langle V_{amp}(t)_{output} \right\rangle = \frac{\eta eG}{h\nu} \; (5,000)(\gamma)(m(t))(P_o)$$

$$\sigma^2 = \left\langle \left[V_{amp}(t)_{output} - \left\langle V_{amp}(t)_{output} \right\rangle \right]^2 \right\rangle$$

$$= \frac{2\eta e^2 G^2 F(G)}{h\nu} \; (5,000)^2 P_o B + \frac{8kTB}{50} \; (5,000)^2 \; , \tag{5.20}$$

where

$$kT = (\text{Boltzmann's constant})(\text{abs. temp}) \sim 4.15 \times 10^{-21}$$

$$B = \text{Bandwidth} = 1 \text{ MHz} = 10^6 .$$

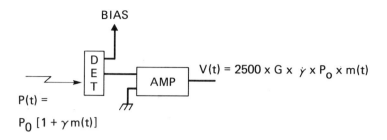

Figure 5.23. Receiver example.

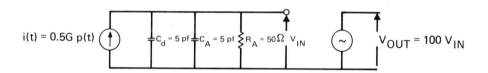

Figure 5.24. Equivalent circuit.

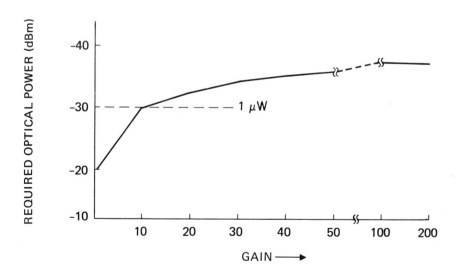

Figure 5.25. Required optical power versus gain.

In Eq. (5.21) we have omitted the dc component of the average output signal.

Now, assume that we wish to have a peak SNR of 10^4. That is,

$$\frac{\left(\frac{\eta eG}{h\nu} (5,000)(\gamma)(P_o)\right)^2}{\sigma^2} = 10^4 . \tag{5.21}$$

If we know γ, G, and $F(G)$ we can solve for the required average optical power P_o. If we assume that the detector ionization ratio is 0.025, we can plot $P_{optical}$ required versus G as in Figure 5.25. We see that the optimal avalanche gain for $\gamma = 0.50$ is approximately 100 and that the required power is -37.5 dBm at optimal gain. We also see that the penalty for not using an avalanche detector $(G = 1)$ is 17.5 dB.

Note too, that the amplifier output voltage is about 0.025 V. It should be emphasized that in this example, two important simplifications were in effect. First, the amplifier was chosen to be a simple 50 Ω input impedance type, which considerably simplified the calculation of the thermal noise at the amplifier output. Furthermore, the performance was specified in terms of the output SNR. This is appropriate for analog modulation systems. However, for digital systems, error rate is the performance criterion. The correlation between error rate and SNR is not always straightforward in optical fiber systems (5.11-5.13).

PROBLEMS FOR CHAPTER 5

1.   In a PN photodiode, the doping level in the P region
            is ρ_p and the doping level in the N region is ρ_n. Assume
            $\rho_p \gg \rho_n$. A reverse bias voltage V is applied, causing
            the regions to the left and right of the junction to
            deplete. Calculate the widths of the depletion regions
            in the N and P materials as a function of V, ρ_p, and ρ_n.

2.   Using Eq. (5.2), calculate and plot Z(f) for C_T = 5 pF
            and R = 50 Ω, 500 Ω, and 5,000 Ω.

3.   In Eq. (5.12), assume $N_{thermal}^2$ is the Johnson noise of
            a 5,000 Ω resistor, and assume B equals 10 MHz. Calcu-
            late the noise equivalent power for η = 1.

4.   In Eq. (5.17), assume the same parameter values as above
            (problem 3). Calculate and plot the noise equivalent
            power versus G for k = 0.03. Calculate and plot P_o
            versus G for SNR = 100 and for SNR = 10^6.

REFERENCES

5.1   H. Melchior, M.B. Fisher, and F. Abrams, "Photodetectors
       for Optical Communication Systems," *Proc. IEEE 58*, October
       1970, pp. 1466-1486.

5.2   L.K. Anderson, M. DiDomenico, and M.B. Fisher, "High Speed
       Photodetectors for Microwave Demodulation of Light,"
       Advances in Microwaves 5, L. Young, Ed. (Academic Press,
       New York, 1970).

5.3   H. Melchior, "Demodulation and Photodetection Techniques,"
       Laser Handbook, F.T. Arecchi and E.D. Schulz-Dubois, Eds.
       (Elsevier, Amsterdam Netherlands, North Holland), pp. 628-
       739.

5.4   T.P. Lee and T. Li, "Photodetectors," Ch. 18 in *Optical
       Fiber Telecommunications*, S.E. Miller and A.G. Chynoweth,
       Eds. (Academic Press, New York, 1979), pp. 593-626.

5.5   D. Schinke, R. Smith, and A. Hartman, "Photodetectors,"
       Ch. 3 in *Semiconductor Devices for Optical Communication*,
       H. Kressel, Ed. (Springer-Verlag, Berlin, 1980), pp. 63-85.

5.6   R.J. McIntrye, "Multiplication Noise in Uniform Avalanche
       Diodes," *IEEE Trans. Electron Devices ED-31*, January 1966,
       pp. 164-168.

5.7   R.J. McIntyre and J. Conradi, "The Distribution of Gains
       in Uniformly Multiplying Avalanche Photodiodes," *IEEE
       Trans. Electron Devices ED-19*, June 1972, pp. 713-718.

5.8   S.D. Personick, "New Results on Avalanche Multiplication
       Statistics with Applications to Optical Detection," *Bell
       Syst. Tech. J. 50*, January 1971, pp. 167-189.

5.9   S.D. Personick, "Statistics of a General Class of Avalanche
       Detectors with Applications to Optical Communication,"
       Bell Syst. Tech. J. 50, December 1971, pp. 3075-3095.

5.10   P.P. Webb, R.J. McIntyre, and J. Conradi, "Properties of
       Avalanche Photodiodes," *RCA Review*, June 1974, pp. 234-278.

5.11   S.D. Personick et al., "A Detailed Comparison of Four
       Approaches to the Calculation of the Sensitivity of Optical
       Fiber System Receivers," *IEEE Trans. Commun. 25*, No. 5,
       May 1977, pp. 541-548.

5.12   P. Balaban, "Statistical Evaluation of the Error Rate of
        the Fiberguide Repeater Using Importance Sampling," *Bell
        Syst. Tech. J. 55*, July 1976.

5.13   W. Hauk et al., "The Calculation of Error Rates for Optical
        Fiber Systems," *IEEE Trans. Commun. 26*, No. 7, July 1978,
        pp. 1119–1126.

5.14   S.D. Personick, "Receiver Design for Digital Fiber Optic
        Communication Systems," *Bell Syst. Tech. J. 50*, No. 1,
        July–August 1973, pp. 843–886.

CHAPTER 6

DESIGN OF RECEIVERS AND TRANSMITTERS FOR FIBER SYSTEMS

S. D. Personick

TRW Technology Research Center
El Segundo, California 90245

6.1   INTRODUCTION

In Chapter 5 we described the process of photodetection —
the conversion of optical power into an electrical current. We
reviewed the physical mechanism, and the mathematical model for
that mechanism both for PIN diodes and for avalanche photodiodes.
In this chapter we describe the theory and practice of receiver
and transmitter design.

6.2   DIGITAL REPEATERS

A typical digital repeater consists of a detector, an
amplifier, an equalizer, and a regenerator, followed by a driver-
optical source pair as shown in Figure 6.1. Once the signal has
been amplified, the remainder of the processing, up to the driver,
is fairly conventional; that is, it is identical with the processing done in conventional wire system repeaters. Although we
shall review the total repeater design, we shall be particularly
interested in the low-noise amplifier that must follow the photodetector and in the driver circuitry for the optical source.
Thus, we begin with a review of noise in baseband amplifiers.

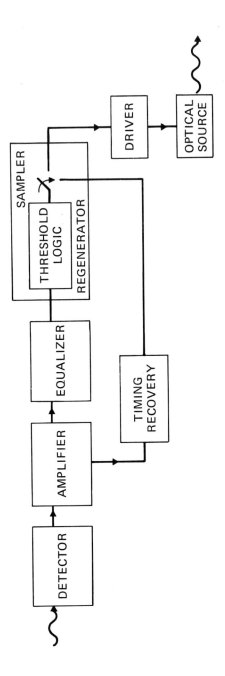

Figure 6.1.   Block diagram of a digital fiber system repeater.

6.3   AMPLIFIERS FOR CAPACITIVE SOURCES

We entitle this section "Amplifiers for Capacitive Sources" because the photodiode equivalent circuit is essentially a current source in parallel with a capacitor. This is shown in Figure 6.2, and follows from the description in Chapter 5. Neglecting shot noise for now, the current $i_s(t)$ is proportional to the optical power falling on the detector. The capacitance is the junction capacitance of the detector, plus "header" and lead capacitance, typically a few picofarads (but conceivably a fraction of a picofarad in small-area photodiodes mounted as "chips" on an amplifier substrate). The proportionality constant relating milliamperes of current to milliwatts of power is the detector responsivity, which is typically 0.5 multiplied by the avalanche gain, if any (for silicon detectors at 0.9 μm wavelength).

Schematically, the amplifier can be modeled as shown in Figure 6.3. Note the following features. The amplifier has an input impedance consisting of a capacitance C_a in parallel with a resistance R_{in}. This resistance may be large or small, depending on the type of device being used. (This point is discussed later.) The capacitance will generally be a few picofarads, although smaller values are possible with special device design. The amplifier gain is represented by the voltage-controlled current source g_m. For field-effect transistors (FETs), g_m is fixed. Typically, it is about 1,000 to 5,000 μS for silicon FETs. For bipolar transistors, g_m is the current gain β divided by the transistor input resistance r_{in}. Field-effect transistor and bipolar examples are shown in Figure 6.4. These are just examples; therefore, one is not constrained to use the common emitter or common source configuration.

Referring back to Figure 6.3, a feedback impedance Z_f is included. This can be stray capacitance, or it can be a lumped component intentionally included by the designer. To limit the

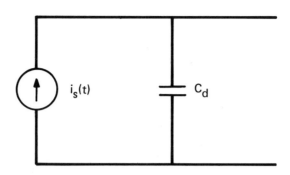

Figure 6.2.   Schematic of a capacitive source.

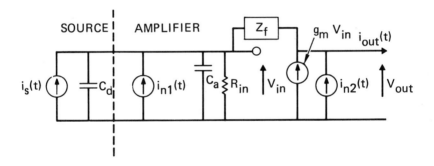

Figure 6.3.   Schematic of a detector/preamplifier.

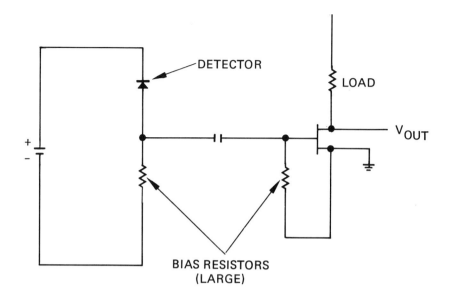

Figure 6.4(a).   FET amplifier.

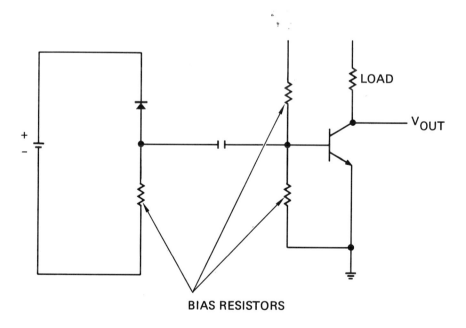

Figure 6.4(b).   Bipolar amplifier.

scope of this discussion, we shall assume that the feedback
impedance is infinite. An alternative approach using an inten-
tional resistive feedback element will be discussed briefly in
Section 6.3.3. (For a more comprehensive study of amplifier
designs, refer to the literature.) We also restrict our discussion here to a single stage of amplification. (Extension of the
ideas to multistage amplifiers is also left to the literature.)

Once again referring to Figure 6.3, we observe two noise
sources. The noise sources are assumed to be independent white
gaussian noise sources with spectral heights I_{n1} and I_{n2} A^2/Hz.
For both the FET and the bipolar transistor, $i_{n1}(t)$ is the shot
noise associated with the gate or base current, respectively,
plus Johnson noise from any biasing resistors at the device input.
In the FET, $i_{n2}(t)$ is associated with the resistance of the source-
drain channel. In the bipolar transistor, $i_{n2}(t)$ is shot noise
associated with the collector bias current.

In designing an amplifier, our goal is to choose a device
and pick an operating point (bias condition) so that the detector
output signal is amplified with as little added noise as possible.
We begin by calculating both the signal and the added amplifier
noise at the amplifier output.

First, we observe that the signal at the amplifier output
is given (in the frequency domain) as follows:

$$I_{out}(\omega) = F[i_{out}(t)] = I_s(\omega) \; \frac{1}{j\omega(C_a + C_d) + \dfrac{1}{R_{in}}} \; g_m \; . \qquad (6.1)$$

Note that the gain versus frequency of the amplifier is
not constant. This is because of the frequency "rolloff" of the
total impedance shunting the signal $i_s(t)$. We could place a
small resistance in shunt with the amplifier input, but that
would add Johnson noise. Instead, we shall include a frequency
"rollup" circuit at the amplifier output as shown in Figure 6.5.
The signal at the output of this "equalizer" is given by

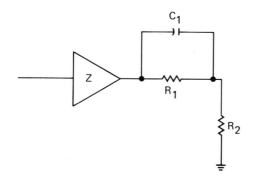

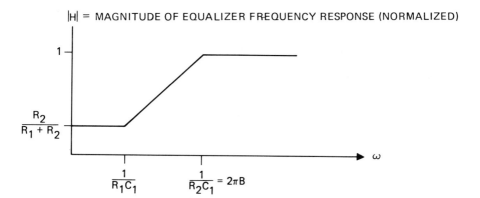

Figure 6.5.   Example of an equalizer circuit.

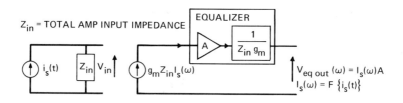

Figure 6.6.   Application of an equalizer circuit.

$$V_{eq \; out}(\omega) = I_s(\omega)A \qquad \text{(for } \omega < 2\pi B) \quad . \qquad (6.2)$$

where A is an arbitrary gain factor. Note that the equalization takes place after further amplification of the signal. The noise spectral density at the equalizer output is given by (Figure 6.6)

$$N_{eq \; out}(\omega) = I_{n1}A^2 + \frac{I_{n2}A^2}{g_m^2}\left[\omega^2(C_a + C_d)^2 + \left(\frac{1}{R_{in}}\right)^2\right] , \qquad (6.3)$$

for $\omega < 2\pi B$.

We see that at each frequency, the noise is minimized if g_m is large, the total capacitance $(C_a + C_d)$ at the amplifier input is small, and the spectral heights I_{n1} and I_{n2} are small. Unfortunately, these parameters are not all independent and at the control of the amplifier designer. To see the tradeoffs, we shall study the FET and the bipolar amplifiers separately.

6.3.1   FET Amplifiers

For FET amplifiers, I_{n1} is typically negligible, at least at bit rates above 0.1 Mb/sec. To simplify our discussion, we shall set it to zero. Also, we shall set R_{in} to infinity, which is also appropriate above 0.1 Mb/sec. In an FET, I_{n2} is given by 2.8 kT g_m, where k is Boltzmann's constant and T is absolute temperature. From Eq. (6.3) we obtain

$$N_{eq \; out}(\omega) = \frac{2.8 \; kT \; A^2\omega^2}{\left[\dfrac{g_m}{(C_a + C_d)^2}\right]} \quad . \qquad (6.4)$$

We see, therefore, that for an FET, $g_m/(C_a + C_d)^2$ is a figure of merit with regard to noise performance. A given FET has a fixed g_m and C_a. However, conceivably, FETs could be designed to optimize this figure of merit. For a given material, e.g., silicon, g_m and C_a trade off so that

$$\frac{g_m}{C_a} \sim \text{constant} \quad . \tag{6.5}$$

Thus, to optimize the figure of merit $g_m/(C_d + C_a)^2$, one should have $C_a = C_d$. In current fiber system art, FETs are not specifically designed to match the detectors. However, efforts have been made to minimize the total capacitance $C_d + C_a$ by minimizing lead lengths and using thin-film hybrid detector-amplifier construction techniques.

It is important to note that the current gain of an FET amplifier is $g_m/\omega(C_a + C_d)$. Thus at high frequencies (typically 25 to 50 MHz) silicon FETs lose their effectiveness as amplifiers. Beyond these frequencies bipolar amplifiers with a controllable transconductance must be used.

6.3.2   Bipolar Amplifiers

Analysis of the bipolar amplifier is somewhat more complicated than the FET analysis. First, we must observe that the value of R_{in} (see Figure 6.3), the amplifier input resistance, is under the control of the designer. Neglecting biasing resistors, R_{in} is given by the small signal transistor input resistance

$$R_{in} = r_{in} = \frac{kT}{eI_{in\ bias}} \quad , \tag{6.6}$$

where $I_{in\ bias}$ is the bias current, and we have neglected ohmic base series resistance.

Next, observe that I_{n1}, the spectral height of the noise source $i_{n1}(t)$, is given by $2eI_{in\ bias} = 2\ kT/R_{in}$. Also, the g_m of the transistor if β/R_{in}, where β is fixed and typically about

100. Finally, the spectral height I_{n2} of noise source i_{n2} is $2e\beta\, I_{in\ bias} = 2\ kT\ \beta/R_{in}$. Substituting into Eq. (6.3) we get

$$N_{eq\ out}(\omega) = \frac{2\ kT}{R_{in}} A^2 + \frac{2\ kT\ R_{in} A^2}{\beta}\left[\omega^2(C_a + C_d)^2 + \frac{1}{R_{in}^2}\right]. \quad (6.7)$$

Since R_{in} is a parameter that we can control, we can optimize Eq. (6.7) to obtain the lowest noise at any particular frequency ω. Also, we can optimize R_{in} to minimize the total noise in some frequency band B. For a given bandwidth B, the optimal R_{in} and minimized noise are given by

$$R_{in\ optimal} = \frac{1}{2\pi(C_a + C_d)B}\ \sqrt{3\beta} \qquad (\text{for } \beta \gg 1)$$

$$N(B)_{eq\ out\ tot} = \frac{1}{2\pi}\int_0^{2\pi B} N_{eq\ out}(\omega)\,d\omega\ \bigg|_{opt\ R_{in}}$$

$$= \left[\frac{4\ kT\ 2\pi(C_a + C_d)}{\sqrt{3\beta}}\right]B^2 A^2. \quad (6.8)$$

From Eq. (6.8) we see that $\beta/(C_d + C_a)^2$ is a figure of merit for bipolar transistors. Note that at optimal bias, the total input impedance shunting the detector rolls off with increasing frequency. Thus, the equalizer does not become superfluous. Also note that if the amplifier is kept at optimal conditions, then the total noise in bandwidth B is proportional to B^2. (For the FET, it is proportional to B^3, as can be derived by integrating Eq. (6.4).)

6.3.3   Example of a Practical Receiver

When designing a practical receiver amplifier, factors
other than minimizing noise can be important. One such factor is
dynamic range. In a typical application, it is undesirable to
require the received optical signal level to be held to a narrow
range near the minimum allowable level. Indeed, it is not unusual
to require the receiver to accommodate ranges of optical input
levels above the minimum by 20 to 40 dB or more.

One disadvantage of a receiver that utilizes an equalizer
to compensate for the amplifier input rolloff, is inadequate
dynamic range. If the amplifier overloads before equalization
has occurred, the signal will be hopelessly distorted.

An alternative amplifier design that has been used in
practical receivers is the transimpedance design. Referring to
Figure 6.7, the feedback element Z_f makes the circuit work as an
operational amplifier with an output voltage proportional to the
product of the detector output current and Z_f. This relationship
holds provided Z_f is not so large that the total gain around the
closed loop falls near or below unity.

The feedback resistor introduces extra input Johnson
noise, and its noise may dominate the total preamplifier noise.
However, this noise disadvantage trades off against a significantly increased dynamic range.

The 50 MHz receiver shown in Figure 6.7 has a 60 dB
electrical dynamic range (corresponding to 30 dB optical level
change capability). It has approximately 10 dB more noise than
an equalizing receiver that can be built with the same components.
This 10 dB of extra noise results in a 5 dB receiver sensitivity
penalty if a PIN detector is used. With a good silicon APD
($k = 0.02$), a sensitivity penalty of only about 1 dB occurs, but
2 to 3 times as much avalanche gain is required (relative to the

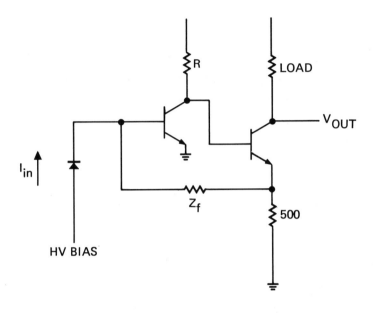

$$V_{out} = Z_f I_{in} \ (LOAD/500)$$

$$R = Z_f = 4000 \, \Omega \text{ for 50 MHz OPERATION}$$

Figure 6.7.   Example of a practical detector/preamplifier circuit.

lower noise amplifier). The requirement for more avalanche gain makes the APD fabrication and yield problems more severe.

Generally, the detector, low-noise amplifier, and a connected fiber pigtail are packaged as a unit in a receiver module. Figure 6.8 shows a plug-in 10 MHz digital receiver "circuit pack" containing such a module. The connected fiber pigtail is clearly visible. This "board" contains additional amplifiers and filters along with an automatic gain control (AGC) circuit as shown in Figure 6.9.

The AGC circuit reduces first the gain of the APD by lowering the high voltage bias, and then reduces the gain of the AGC amplifier, to keep the peak output signal constant in the presence of varying optical input levels. By varying the APD gain over a range of 5 (50 to 10) and by varying the AGC amplifier gain over a range of 1,000 to 1, 37 dB of optical signal level change can be accommodated without overload.

6.3.4   Summary

We studied amplifiers and amplifier noise with the goal of designing an amplifier that adds as little noise as possible to the detector output signal. We later showed that in practical systems, we sometimes trade off increased amplifier noise for other desirable features such as large dynamic range (or convenience of using existing commercial amplifiers). We showed that a frequency rollup equalizer can be used to compensate for the frequency rolloff at the amplifier input (because of the capacitive impedance). We also showed that other approaches, feedback, in particular, can be used to achieve similar results. In the literature, such feedback amplifiers are referred to as "transimpedance" amplifiers. The interested reader is referred to the growing literature on this subject describing both theory and practice in amplifier design.

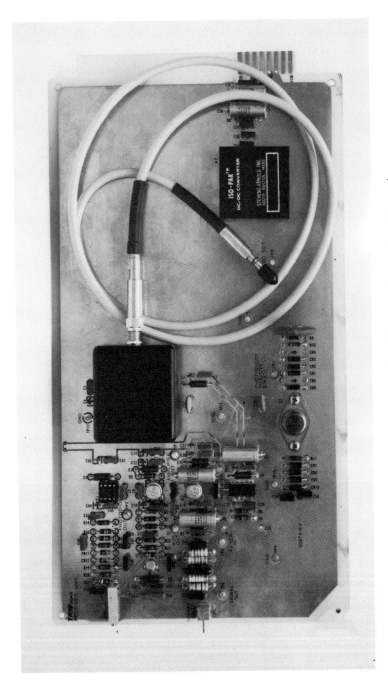

Figure 6.8.   Photograph of a receiver circuit pack.

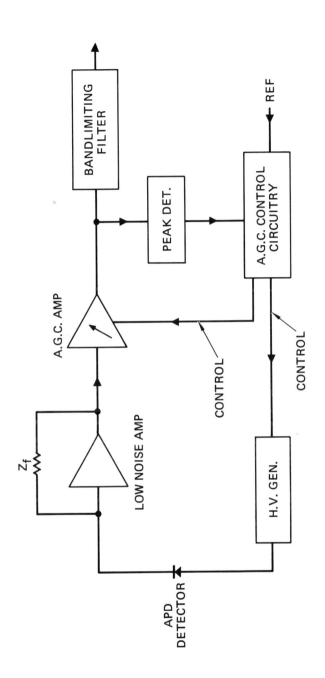

Figure 6.9.   Block diagram of a practical receiver.

6.4   THE LINEAR CHANNEL OF A DIGITAL REPEATER

In Chapter 5, we showed that over a reasonable (but finite) range of optical illumination, the current that flows in the detector, on the average, is proportional to the optical power impinging on it. Figure 6.10 shows what may be called the linear channel of a repeater. As in the detector, the output of each block is a linearly filtered version on the input of that block. Neglecting noise, we see that the linear channel serves two purposes: amplification and equalization.

6.4.1   Equalization for Digital Repeaters

In a digital fiber system, the light falling on the detector is assumed to be a sum of pulses of optical power

$$P_{in}(t) = \Sigma_n a_n h_p(t-nT) \text{ (watts)} \quad , \quad (6.9)$$

where a_n = 0 of 1 for each n, and T is the spacing between pulses.
In Eq. (6.9), the individual pulses may overlap as a result of spreading as the light propagates along the fiber. After detection and amplification, we obtain

$$V_{eq \; out}(t) = \Sigma_n a_n h_{eq \; out}(t-nT) \text{ (volts)} \quad , \quad (6.10)$$

where $h_{eq \; out}(t)$ includes the amplifier equalizer, which compensates for any integration at the amplifier input. The pulses $h_{eq \; out}(t)$ may overlap just as the input power pulses may. This is referred to as intersymbol interference. If $V_{eq \; out}(t)$ were sampled once for each time slot, then the resulting sample values would depend on more than one digit (and noise). If the shape $h_{eq \; out}(t)$ is known (which requires knowledge of $h_p(t)$), then an

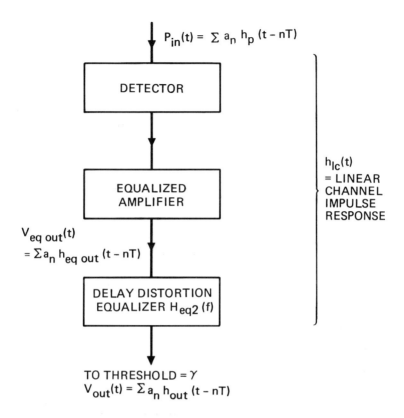

$$P_{in}(t) = \sum a_n h_p (t - nT)$$

DETECTOR

EQUALIZED
AMPLIFIER

$V_{eq\ out}(t)$
$= \sum a_n h_{eq\ out} (t - nT)$

DELAY DISTORTION
EQUALIZER $H_{eq2} (f)$

$h_{lc}(t)$
= LINEAR
CHANNEL
IMPULSE
RESPONSE

TO THRESHOLD = γ
$V_{out}(t) = \sum a_n h_{out} (t - nT)$

Figure 6.10.   Block diagram of a typical linear channel.

equalizer filter can be built to reduce the pulse overlap. If the desired output pulses are $h_{out}(t)$ and the equalizer input pulses are $h_{eq\ out}(t)$, then the equalizer frequency response must be

$$H_{eq\ 2}(f) = \frac{F\{h_{out}(t)\}}{F\{h_{eq\ out}(t)\}} \quad . \tag{6.11}$$

Since $|H_{eq\ 2}(t)|$ is, in general, an increasing function of f (for large initial pulse overlap), noise enhancement occurs. That is, noise entering this second equalizer is increased in mean square value. Thus, there is some tradeoff between reduced intersymbol interference and enhanced noise. As a practical matter, it is tempting to leave out the equalizer ($H_{eq\ 2}$) and accept the intersymbol interference, since inclusion of this equalizer may require more information about the fiber impulse response than is available to the repeater designer.

For example, suppose the received optical pulses are gaussian in shape (as shown in Figure 6.11):

$$h_p(t) = \frac{1}{\sqrt{2\pi(0.25\ T)^2}}\ e^{-t^2/2(0.25\ T)^2}\ , \qquad (6.12)$$

where T is the width of a time slot. Suppose next that after passing through the amplifier we obtain the following individual pulse shape (see Figure 6.11):

$$h_{eq\ out}(t) = \frac{1}{\sqrt{2\pi(0.5\ T)^2}}\ e^{-t^2/2(0.5\ T)^2}\ . \qquad (6.13)$$

The pulses given in Eq. (6.13) overlap. As a measure of inter-symbol interference, we can plot what is referred to as an eye diagram. It consists of two curves. The first is a plot of the largest value $v_{out}(t)$ can take when $a_o = 0$ and the other a_n can be chosen arbitrarily (see Eq. (6.10) and Figure 6.10). The second curve is a plot of the smallest value $v_{out}(t)$ can assume

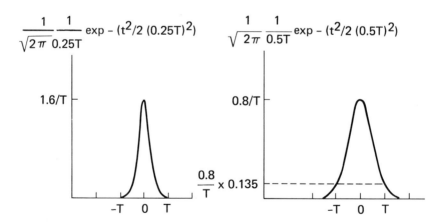

Figure 6.11.   Illustration of intersymbol interference,−pulses.

if a_o = 1 and the other a_n are again chosen arbitrarily. The eye
is plotted in Figure 6.12 for the pulse shape of Eq. (6.13).
Referring to Figure 6.12, the value d = peak eye opening/individual
pulse height is called the fractional eye opening. It is a measure of how much the ability to distinguish between a_o = 0 and
a_o = 1 has been degraded.

Once again, in principle, we could build an equalizer to
modify the individual pulse shape resulting in a larger fractional eye opening. However, this enhances the noise, producing a
tradeoff.

6.5   CALCULATING THE ERROR RATE OF A·DIGITAL REPEATER

Referring to Figure 6.10, we see that the decision process consists of sampling the linear channel output once for each
time slot. If the sample value exceeds some threshold, we decide
that a_n = 1 (see Eq. (6.9)); if the sample is below the threshold,
we decide that a_n = 0. The probability of error is the chance
that $v_{out}(nT)$ exceeds threshold and a_n = 0, or $v_{out}(nT)$ is below
threshold and a_n = 1. To calculate this error probability, we
would have to know the probability distribution of $v_{out}(t)$ under
both hypotheses (a_n = 0, a_n = 1). In receivers with avalanche
gain, it is difficult to make this calculation exactly. Various
approximation techniques are available. The simplest is the
gaussian approximation, which is outlined next. (The interested
reader is referred to the evolving literature for more detailed
information and alternative approximations.)

6.5.1   The Gaussian Approximation

Consider the output of the linear channel shown in
Figure 6.10. At sampling time nT, this output is given by

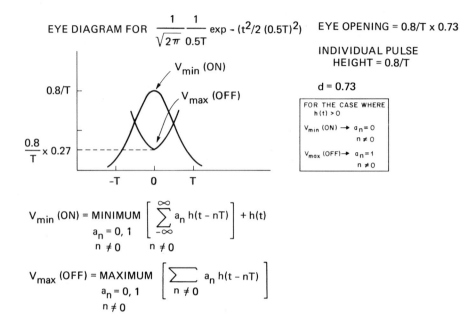

$$V_{min} (ON) = \underset{\substack{a_n = 0, 1 \\ n \neq 0}}{MINIMUM} \left[\sum_{\substack{-\infty \\ n \neq 0}}^{\infty} a_n h(t - nT) \right] + h(t)$$

$$V_{max} (OFF) = \underset{\substack{a_n = 0, 1 \\ n \neq 0}}{MAXIMUM} \left[\sum_{n \neq 0} a_n h(t - nT) \right]$$

Figure 6.12.   Illustration of intersymbol interference,-eye diagram.

$$v_{out}(t)\Big|_{t=nT} = \Sigma_j a_j h_{out}(t-jT) + N(t)\Big|_{t=nT} \quad , \qquad (6.14)$$

where the noise $N(t)$ consists of two parts: amplifier thermal noise and randomly multiplied shot noise associated with the detection process. (The thermal noise was discussed in Section 6.3. The randomly multiplied shot noise was discussed in Chapter 5.)

Recall that the detector output consists of electrons that are created at random times, but at an average rate proportional to the incident optical power. The average current emitted by the detector is thus proportional to the incident optical power. The deviations from this average current (caused by the randomness of when and how many electrons will be emitted) results in the randomly multiplied shot noise. In the gaussian approximation, we pretend that $N(nT)$ is a gaussian random variable. With this approximation, we need only know the mean and variance of $N(nT)$. The thermal noise has zero mean value, and the randomly multiplied shot noise has mean value also equal to zero (the mean value of the current is included in the signal term in Eq. (6.14). Thus $N(nT)$ is completely characterized by its variance under this approximation. We obtain the following two expressions from Figure 6.10, and the statistical model of a Poisson process discussed in Chapter 5.

$$\left\langle v_{out}(nT) \right\rangle = v_{out\ avg}(nt) = \frac{en\bar{G}}{h\nu} \int P_{in}(t')h_{1c}(t-t')dt'$$

$$\Big|_{t=nT} = \Sigma a_j h_{out}(t-jT)\Big|_{t=nT}$$

$$\left\langle N^2(nT) \right\rangle = N_{avg}^2(nT) \tag{6.15}$$

$$= N_{amp}^2 + \int P_{in}(t')e^2\frac{\eta}{h\nu}\overline{G^2}h_{lc}^2(t-t')dt'\Big|_{t=nT}$$

$$= \frac{e^2\eta\overline{G^2}}{h\nu}\Sigma a_j\int h_p(t'-jT)h_{lc}^2(t-t')dt'\Big|_{t=nT} + N_{amp}^2 \quad ,$$

where

$\quad e\eta/h\nu$ = detector responsivity, not including avalanche gain

$\qquad e$ = electron charge

$\qquad \overline{G}$ = average avalanche gain

$\qquad \overline{G^2}$ = mean square avalanche gain

$\quad h_{lc}(t)$ = linear channel impulse response

$\qquad N_{amp}^2$ = amplifier contribution to output noise.

From Eq. (6.15) we see that, in general, both the signal $v_{out\ avg}(nT)$ and the noise $N_{avg}^2(nT)$ depend on the complete set of symbol values (a_j). As a practical matter, this dependence is very weak for $|j-n|$ larger than some small integer. Thus, it is possible to calculate the signal and noise variance for all possible symbol combinations that affect these two quantities. For each combination of $\{a_j\}$ for $j \neq n$, the probabilities of error are given by

$$P_f = prob(v_{out}(nT) > threshold)$$

$$if \ a_n = 0, \ \{a_j\} \ given \ for \ j \neq n$$

$$P_m = prob(v_{out}(nT) < threshold)$$

$$if \ a_n = 1, \ \{a_j\} \ given \ for \ j \neq n \ . \tag{6.16}$$

Since we have assumed that $N(nT)$ is approximately a gaussian random variable, these probabilities reduce to

$$P_f = \text{ERFC*} \left[\frac{\gamma - [v_{out\ avg}(nT) | \{a_j\}, a_n = 0]}{\sqrt{(N^2_{avg}(nT) | \{a_j\}, a_n = 0}} \right]$$

$$P_m = \text{ERFC*} \left[\frac{[v_{out\ avg}(nT) | \{a_j\}, a_n = 1] - \gamma}{\sqrt{N^2_{avg}(nT) | \{a_j\}, a_n = 1}} \right]$$

(6.17)

where

$$[x | \{a_j\}, a_n] \triangleq \text{ particular value } x \text{ assumes for the}$$
given set of $\{a_j\}$ and a_n, $\gamma \triangleq$ threshold for decisions,

and

$$\text{ERFC*} (x) = \frac{1}{\sqrt{2\pi}} \int_x^\infty e^{-y^2/2} dy \quad . \tag{6.18}$$

ERFC* is a tabulated function and is shown in Figure 6.13. Using Eq. (6.17), we can find the optimal threshold, avalanche gain, and required optical power to achieve a desired error rate. We can calculate the effects of pulse spreading in the fiber on the receiver sensitivity (required optical power to achieve a desired error rate). Note, however, that the gaussian approximation is not completely satisfactory in systems using avalanche detectors. (Consult the literature for more tedious, but more accurate analysis techniques.) Laboratory experience shows that the gaussian approximation is adequate for the calculation of receiver sensitivity (to within ~ 1 dB). In practice, one shortcoming is

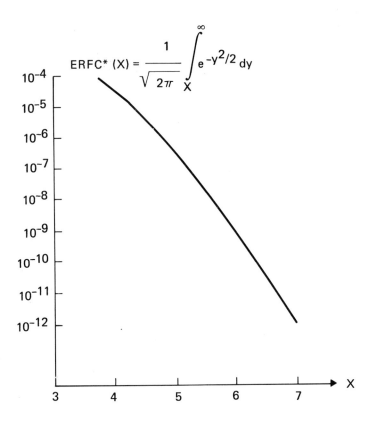

Figure 6.13. Plot of the error function.

that in systems with avalanche gain, the optimal threshold is closer to the center of the "eye" than the gaussian approximation would predict. This is a consequence of the highly skewed nature of the probability distribution of randomly multiplied shot noise.

6.6   COMPUTATIONAL RESULTS ON ERROR RATE OF A DIGITAL
         REPEATER

        Based on the gaussian approximation, receiver sensitivity curves have been calculated for receivers that use bipolar or FET front ends and for receivers with and without avalanche gain. Assuming no significant pulse spreading in the fiber, the results are shown in Figure 6.14. In that figure, a 10^{-9} error rate is assumed. The rather large curve widths take into account component ranges. Crosses on the curves are for experimental systems that have been reported. For example, at 45 Mb/sec, a receiver sensitivity of -57 dBm of optical power was achieved. Thus, for a transmitted power level of 1 mW, 57 dB for fiber loss, connector loss, and margin could be allowed, provided pulse spreading is sufficiently small.

The effects of pulse spreading have also been calculated for typical systems. It has been found that pulse spreading is negligible if the impulse response of the fiber $h_{fiber}(t)$ satisfies

$$\sigma < 0.25 \ T = 0.25 \ \text{pulse spacing} \quad , \quad (6.19)$$

where

$$\sigma^2 = \frac{\int t^2 h_{fiber}(t) \, dt}{\int h_{fiber}(t) \, dt} - \left[\frac{\int t h_{fiber}(t) \, dt}{\int h_{fiber}(t) \, dt} \right]^2 . \quad (6.20)$$

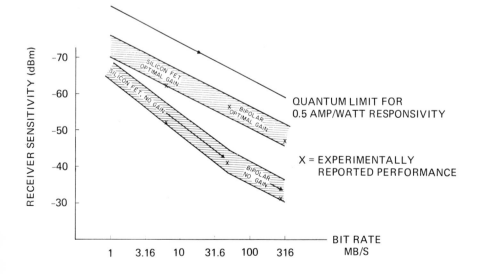

Figure 6.14. Curves of typical receiver performance.

If the fiber impulse response satisfies Eq. (6.19), then the
allowable fiber loss is the difference between the transmitted
power and the receiver sensitivity. Under those circumstances,
the receiver is said to be loss limited. If the fiber pulse
spreading is excessive, then the fiber length must be smaller
than the loss-limited value to reduce the intersymbol interference. The receiver is then said to be dispersion limited.

6.7   TIMING RECOVERY FOR DIGITAL REPEATERS

        Timing recovery for sampling the output of the linear
channel is accomplished in the same way as for wire-system repeaters. Either a filter or a phase-locked loop can be used.
One interesting difference between the fiber system and the wire
system is that the signal in the linear channel may contain a
timing component even without rectification. This is because
the optical system uses "on-off" signaling rather than plus-
minus signaling as is used in wire systems. For this timing component to be present, the transmitted signal must be in a "return-
to-zero" format. If desired, the linear channel signal can be
differentiated and rectified as in a wire repeater in order to
obtain an enhanced timing component or if a non-zero-to-return
format is used at the transmitter.

6.8   DRIVER CIRCUITRY

        Optical sources used in fiber systems are high-current
low-impedance devices. Typical operating currents exceed 100 mA.
Typical impedances are a few ohms. Thus, for efficient operation, high-current low-impedance transistor drivers are needed.
At modest modulation speeds (less than 25 Mb/sec) the optical
source will follow the modulation (in a nonlinear fashion). At
high modulation speeds, the optical components exhibit

pattern-dependent effects, where the height of one emitted pulse depends on whether pulses were emitted in previous time slots. Thus, at high modulation rates, fairly sophisticated driver designs are required. Successful modulation of light-emitting diodes (LEDs) at up to 100 Mb/sec and lasers beyond 1000 Mb/sec have been reported.

When laser sources are used, one generally biases the device close to the threshold current, using a relatively small (10 to 20 mA) incremental signal for modulation. Compensation must be included in the driver for the temperature-dependent threshold of these devices. The threshold may also change as the device degrades with age. One approach that has been used is to monitor the light emitted by the laser locally with a PIN detector. The detector output is then used in a feedback arrangement to stabilize the output of the laser. This feedback approach has also been used to linearize LEDs for analog modulation purposes.

Figure 6.15 shows a typical transistor-transistor logic (TTL) compatible LED driver circuit. Figure 6.16 shows an emitter-coupled logic (ECL) compatible balanced driver which provides faster switching and reduced power supply noise. Figure 6.17 shows a feedback scheme for stabilization of the output of a laser transmitter. Figure 6.18 shows a transmitter module including an ECL-compatible LED driver and a connected output fiber pigtail.

6.9   ALTERNATIVE MODULATION FORMATS

The preceding discussion has been restricted to digital modulation, since that appears most promising for optical transmission in the near future. Other modulation formats are possible and have been reported in the literature.

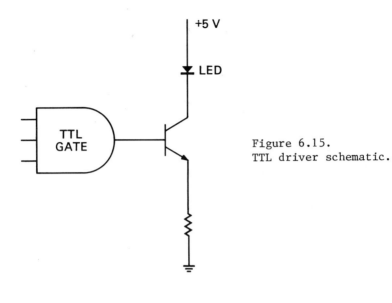

Figure 6.15.
TTL driver schematic.

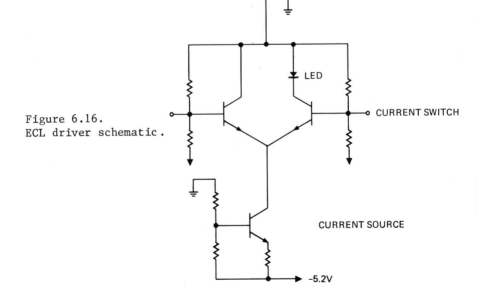

Figure 6.16.
ECL driver schematic.

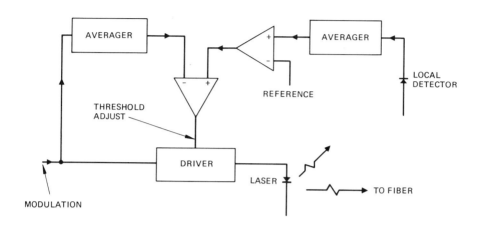

Figure 6.17.   Feedback stabilized driver block diagram.

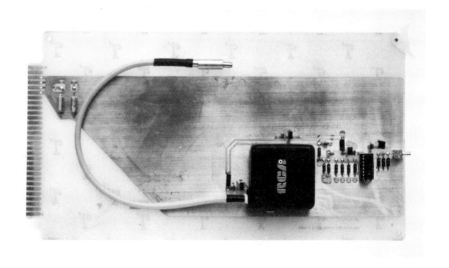

Figure 6-18.   Photograph of a transmitter circuit pack.

If linearity is not too severe a problem, direct analog modulation of LEDs (or perhaps even lasers) is possible. In such a system, the LED drive current is linearly modulated about some operating point. Optical feedback can be used to improve the linearity.

Pulse-position modulation may be desirable for use with laser sources that emit narrow pulses of high-peak power (but which must operate at low duty cycles). In such systems, linearity depends on the ability to linearly delay the emission time of a pulse.

Subcarrier frequency modulation has also been proposed for fiber systems. Here, however, the delay distortion in the fiber may be a severe limitation. In subcarrier systems, a sinusoidal drive current is superimposed upon a bias to form the LED input. The sinusoidal component is then modulated using any FM format.

6.10   CONCLUSIONS

Sensitive receivers for optical fiber systems have been designed. Allowable losses between transmitter and receiver of 50 dB or more are possible at data rates beyond 100 Mb/sec. For best performance, the receiver front-end amplifier must be designed to work with a capacitive source, such as a PIN or avalanche photodiode. The driver must be designed to work efficiently with a high-current, low-impedance optical source. Beyond this, however, much of the optical repeater circuitry is identical with that of a wire repeater.

PROBLEMS FOR CHAPTER 6

1. Consider a detector-amplifier as shown in Figure 6.3, with
    C_d = 5 pF and C_a = 5 pF. Calculate and plot Z_{in} versus f
    for R_{in} = 50 Ω and R_{in} = 5,000 Ω.

2. Consider the amplifier equalizer shown in Figures 6.5 and
    6.6. Assume Z_{in} consists of a 10 pF capacitor in parallel
    with a 10 kΩ resistor. Calculate the values R_1 and C_1
    required to obtain an equalized response which rolls off at
    20 MHz (3 dB down) assuming R_2 = 50 Ω. Plot the equalized
    response.

3. For an FET amplifier as shown in Figure 6.5, assume
    g_m = 5,000 μS, Z = 10^6, and C_a = C_d = 5 pF. Assume ideal
    bandlimiting and equalization to 10 MHz (infinitely sharp
    cutoff). Calculate the rms output noise voltage.

4. Assume the parameters of problem 3. If an impulse of charge
    1.6×10^{-19} C (1 electron) is emitted by the detector, cal-
    culate the peak amplifier-equalizer response in volts.
    Calculate the ratio of this peak response (to a single elec-
    tron) to the rms output noise.

5. For the driver shown in Figure 6.15, assume that the TTL gate
    output voltage is 5 V in the "on" state. Assume that the
    base-emitter voltage drop of the transistor is 0.7 V. Calcu-
    late and plot the required value of the emitter-to-ground
    resistor versus LED current in the "on" state for currents
    between 10 to 100 mA. Neglect transistor emitter resistance.

REFERENCES

6.1   A Van der Ziel, *Noise: Sources, Characterization*
       (Prentice Hall, Englewood Cliffs, New Jersey, 1970).

6.2   A.B. Gillespie, *Signal, Noise, and Resolution in Nuclear
       Particle Counters* (Pergamon Press, Inc., New York, 1953).

6.3   S.D. Personick, "Receiver Design for Digital Fiber Optic
       Communication Systems," *Bell Syst. Tech. J. 50*, No. 1,
       July–August, 1973, pp. 843–886.

6.4   J.E. Goell, "Input Amplifiers for Optical PCM Receivers,"
       Bell Syst. Tech. J. 53, No. 9, November 1974, pp. 1711–1794.

6.5   *Semiconductor Devices for Optical Communication*, H. Kressel,
       Editor (Springer Verlag, 1980), Ch. 4.

6.6   J.E. Goell, "An Optical Repeater with a High Impedance
       Input Amplifier," *Bell Syst. Tech. J. 53*, No. 4, April 1974,
       pp. 629–643.

6.7   P.K. Runge, "An Experimental 50 Mb/s Fiber Optic Repeater,"
       IEEE Trans. Commun. 24, April 1976, pp. 413–418. Also,
       "A 50 Mb/s Repeater for a Fiber Optic PCM Experiment,"
       Proc. ICC, Minneapolis, 1974.

6.8   J.E. Goell, "A 275 Mb/s Optical Repeater Experiment
       Employing a GaAs Laser," *Proc. IEEE (Letters) 61*,
       October, 1973, pp. 1504–1505.

6.9   T. Ozeki and T. Ito, "A 200 Mb/s PCM DH GaAlAs Laser
       Communication Experiment," presented at 1973 Conf. Laser
       Engineering and Applications, Vol. *QE-9*, Washington, D.C.,
       June 1973, p. 692.

6.10   J.L. Hullett, and T.V. Moui, "A Feedback Receive Amplifier
       for Optical Transmission Systems," *IEEE Trans. Commun. 24*,
       October 1976, pp. 1180–1185.

6.11   F.M. Banks et al., "An Experimental 45 Mb/s Digital Trans-
       mission System Using Optical Fibers," *Proc. ICC*,
       Minneapolis, 1974.

6.12   T. Uchida et al., "An Experimental 123 Mb/s Fiber Optic
       Communication System," *Proc. Topical Meeting on Optical
       Fiber Transmission*, Williamsburg, Virginia, January 7–9,
       1975.

6.13. D. Sell and T. Maione, "Experimental Fiber Optic Transmission System for Interoffice Trunks," *IEEE Trans.
Commun.* 25, No. 5, May 1977, pp. 517-522.

6.14 R.W. Lucky, J. Salz, and E. Weldon, *Principles of Data
Communication* (McGraw-Hill, New York, 1968).

6.15 W. Davenport and W. Root, *Random Signals and Noise*
(McGraw-Hill, New York, 1958).

6.16 M. Chown, et al., "Direct Modulation of Double Heterostructure Lasers at Rates up to 1 Gb/s," *Electron. Lett.*
9, January 1973, pp. 34-36.

6.17 P.K. Runge, "An Experimental 50 Mb/s Fiber Optic Repeater,"
IEEE Trans. Commun. 24, April 1976, pp. 413-418. Also,
"A 50 Mb/s Repeater for a Fiber Optic PCM Experiment,"
Proc. ICC, Minneapolis, 1974.

6.18 *Semiconductor Devices for Optical Communication,*
H. Kressel, Editor (Springer Verlag, 1980), Ch. 5.

6.19 W.S. Holden, "Pulse Position Modulation Experiment for
Optical Fiber Transmission," *Proc. Topical Meeting on
Optical Fiber Transmission*, Williamsburg, Virginia,
January 7-9, 1975.

6.20 A. Szanto and J. Taylor, "An Optical Fiber System for
Wideband Transmission," *Proc. ICC*, Minneapolis, 1974.

6.21 A. Albanese, H. Lensing, "I.F. Lightwave Entrance Links
for Satellite Communications," *Proc. ICC*, Boston, Mass.,
June 1979, Ch. 1435-7 CSCB.

6.22 S.D. Personick, "Receiver Design for Digital Fiber Optic
Communication Systems," *Bell Syst. Tech. J. 50*, No. 1,
July-August, 1973, pp. 843-886.

6.23 S.D. Personick, "Statistics of a General Class of Avalanche
Detectors with Applications to Optical Communication,"
Bell Syst. Tech. J. 50, No. 10, December 1971, pp. 3075-
3095.

6.24 S.D. Personick, "New Results on Avalanche Multiplication
Statistics with Applications to Optical Detection,"
Bell Syst. Tech. J. 50, No. 1, January 1971, pp. 167-189.

6.25 S.D. Personick, "Receiver Design for Optical Fiber
Systems," *Proc. IEEE 65*, No. 12, December 1977, pp. 1670-
1678.

6.26   S.D. Personick et al., "A Detailed Comparison of Four
        Approaches to the Calculation of the Sensitivity of Opti-
        cal Fiber System Receivers," *IEEE Trans. Commun. 25*,
        May 1977, pp. 541–548.

6.27   G.L. Cariolaro, "Error Probability in Digital Fiber Optic
        Communication Systems," *IEEE Trans. Inf. Theory IT-24*,
        March 1978.

6.28   W. Hauk et al., "The Calculation of Error Rates for Optical
        Fiber Systems," *IEEE Trans. Commun. 26*, No. 7, July 1978,
        pp. 1119–1126.

6.29   M. Mansuripur et al., "Fiber Optics Receiver Error Rate
        Prediction Using Gram-Charlier Series," *IEEE Trans.
        Commun. 28*, No. 3, March 1980, pp. 401–407.

NOTE: References 6.1 through 6.6 refer to Aplifier Noise.

        References 6.7 through 6.13 refer to Repeater Design.

        References 6.14 and 6.15 refer to Statistical Communication

        Theory.

        References 6.16 through 6.18 refer to Driver Circuits.

        References 6.19 through 6.21 refer to Analog Systems.

        References 6.22 through 6.29 refer to Error Rate Calculations.

CHAPTER 7

DESIGN CONSIDERATIONS FOR MULTITERMINAL NETWORKS

M.K. Barnoski

TRW Technology Research Center
El Segundo, California 90245

7.1   INTRODUCTION

The principal intent of this chapter is to utilize the
information presented in the preceding chapters of this text to
address the problems in applying fiber optic transmission lines
to distribute information to numbers of remote terminals, all
interconnected in a bidirectional data distribution system. Such
data distribution networks are of particular interest in the
short-range class of applications which include information
transfer within vehicles such as aircraft and ships and also
within structures such as office buildings, manufacturing facil-
ities, power plants, and power switching yards. For these types
of applications, maximum interterminal spacings will be a few
hundred meters, with information rates ranging from kilohertz to
possibly a few hundred megahertz.

In addition to these relatively short-range applications, it is also interesting to consider the more long-range
possibilities such as the utilization of fiber-optic technology
for distribution of data in a broadband network. Both of these
applications utilize the concept of the data bus, a single transmission line which simultaneously carries many different multi-
plexed signals and serves a number of spatially distributed

terminals. The engineering design considerations that must be
investigated to properly design multiterminal fiber-optic networks will now be discussed.

The components that comprise a typical terminal-to-
terminal section of a multiterminal system are illustrated schematically in Figure 7.1. These are the transmitting LED or laser
source, the input and output couplers, the fiber cabling, the
other distribution system components (which from a terminal-to-
terminal viewpoint represent added loss), and, finally, the receiving photodetector. Clearly, the amount of optical power
available for distribution in the system depends on the amount of
power launched into the transmission line at any given terminal,
the amount lost in the fiber waveguide cabling as a result of
attenuation, and the amount of optical power required at the receiving photodiode necessary to maintain the error rate (signal-
to-noise ratio) desired. Thus, the design of a data distribution
system dictates that the following set of questions be addressed:
(1) What error rate is required? (2) How much optical power is
required incident on the photodetector to maintain the desired
error rate? That is, how good is the receiver? (3) How much
optical power is emitted from the source? (4) How much optical
power is coupled into the transmission line at the input? (5)
How much optical power is available for data distribution? That
is, how high can the distribution system losses be? (6) What
is the most efficient distribution system for your particular
application?

7.2   MINIMUM DETECTABLE RECEIVER POWER

Assuming that the acceptable error rate is one error
for every 10^9 bits transmitted, the minimum detectable power required incident on the active surface of the photodetector can be
determined as discussed in Chapter 6. For convenient reference,

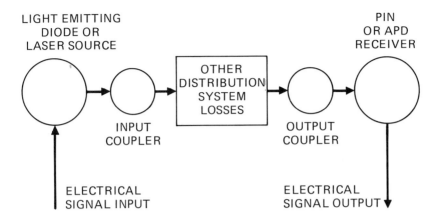

Figure 7.1   Components comprising a terminal-to-terminal section
of a multiterminal distribution system.

plots of minimum detectable power (7.1) versus information bit
rate are reproduced in Figure 7.2. As can be seen, the minimum
power incident on the receiver necessary to maintain an error
rate of 10^{-9} increases with increasing information bit rate.
This is true for both silicon PIN (no gain) and APD (avalanche
photodiode with optimal gain) detectors with either FET or bipolar front end preamplifiers. Plots were made for both bipolar
and FET input amplifiers. The shaded areas account for component
variations. The data points shown are experimental results obtained at various bit rates. From the figure it can be seen that,
for example, at 50 Mbit/sec data rate, the power required at a
receiver using a PIN photodiode is approximately -42 dBm, while
that required if an APD is used is only -57 dBm. These results
are valid for the case where the optical pulse width is less than
a time slot, e.g., in systems where intersymbol interference is
negligible. This is a valid assumption for relatively short-
length, medium-data-rate distribution systems.

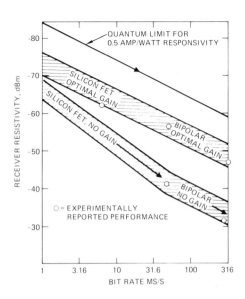

Figure 7.2   Receiver sensitivity versus bit rate.

7.3   COMPATIBLE LIGHT-EMITTING SOURCES

Given the optical power required incident on the
receiver, it is necessary next to determine the amount of power
coupled into the transmission line at the input. This, of course,
depends on the total power emitted from the source. As mentioned
in Chapter 3, the two principal opto-electronic sources of interest for use in fiber optic communications systems are the LED and
the injection-laser diode. The total optical output power emitted
from a high-brightness LED and from a cw room-temperature injection laser is plotted as a function of drive current in Figure 7.3.

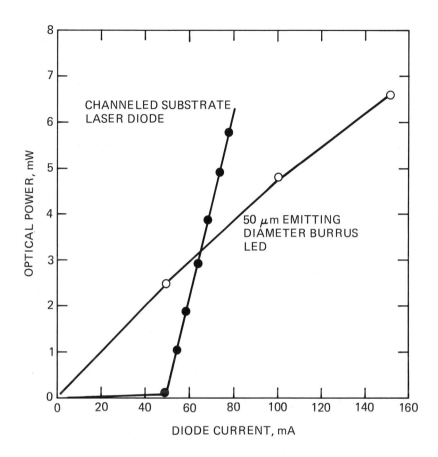

Figure 7.3   Total optical output power from a Burrus type LED
and a channeled substrate laser as a function of
diode forward current.

The power-versus-current plots shown in the figure,
which are typical of devices available today, establish the total
amount of optical power available from the source.   The fraction
of this power coupled into the fiber transmission line at the input must be determined.   The problem of determining the input
coupling efficiency is treated in detail in Chapter 3.   The input
coupling efficiency for a surface-emitting LED was shown to be
NA^2, which is -14 dB for a 0.2 NA fiber.   Since the injection

laser has a more directional radiation pattern, the coupling
efficiency is greater than that for the LED. Typical coupling
loss for an injection-laser diode into a 0.2 NA fiber is about
-6 dB (refer to Figure 4.19 of Chapter 4). For the purpose of
illustration, assume that the injection-laser diode with the
power-current curve shown in Figure 7.3 is used as a source.
Also assume that the data stream to be transmitted is a sequence
of "zeros" and "ones" and that the average current level of the
data stream is 70 mA. In this case, the average optical power
emitted by the laser is approximately 6.3 dbm (4.3 mW). The average power coupled into a 0.2 NA fiber is therefore approximately
0 dBm. If, however, the LED with the power current curve shown in
Figure 7.3 is used as the source, the average power emitted by
the device is 5.3 dBm (3.4 mW). In this case, the average optical power coupled into the 0.2 NA fiber is approximately -9 dBm.

7.4   DISTRIBUTION SYSTEMS COMPONENTS

Current trends in system design strongly favor micro-
miniaturization, digital processing, and system level integration
in order to achieve smaller size and weight, consume less power,
lower costs, and improve reliability. These trends naturally
point to data bus multiplexing, i.e., the interconnection of a
number of spatially distributed terminals via fiber-optic waveguide cables. Two basic fiber-optic configurations are currently
being considered for the distribution of data to a set of remote
terminals: a serial distribution system that uses access (T)
couplers and a parallel system employing a star coupler.

Optical access couplers such as the "T" and "star"
are important in the design of fiber-optic bus systems. Much
effort has been expended in the investigation of various techniques for fabricating these devices. A particularly promising
approach for both T and star couplers is the fused-fiber
coupler (7.2,7.3). In this coupler, the individual fiber

waveguides are thermally fused and pulled into a twin biconical
taper. The coupling efficiency is controlled by controlling the
core-to-core spacing and taper length. A schematic of two fused
fibers with a focused He-Ne laser exciting the main channel is
shown in Figure 7.4. The optical radiation emanating from the
output ends of the two coupled fibers is displayed by directly
exposing a photographic film located as shown in the figure. A
photomicrograph of a cleaved cross section of the two fused
fibers is shown in Figure 7.5. The two cores and surrounding
cladding regions are clearly visible in the figure. Data sum-
marizing the performance of 36 couplers are shown in Figure 7.6.
The coupling ratio is defined as the coupled power divided by
the total output power $P_4/(P_3+P_4)$; the excess loss is defined as
the total output power divided by the total input power
$(P_3+P_4)/P_1$. In the figure the average loss is plotted versus
average coupling ratio for each of the 36 couplers. The design
goal for these devices was 12 each of 3, 6, and 10 dB taps, each
with less than 1 dB excess loss.

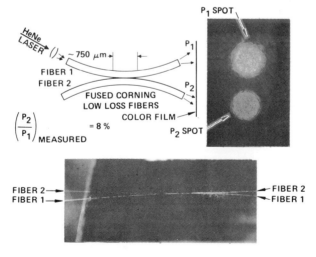

Figure 7.4   Schematic of the two fused
              fibers with a fucused He-Ne
              laser exciting the main
              channel fiber.

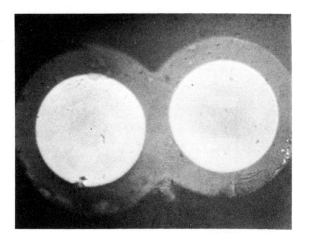

Figure 7.5   Photomicrograph of a cleaved
cross section of the two
welded fibers.

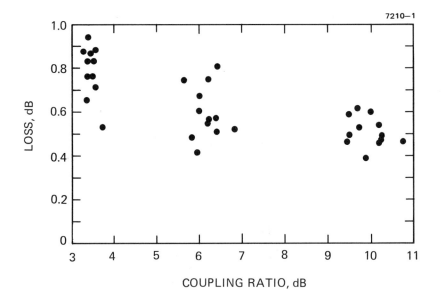

Figure 7.6   T-coupler data.   Each point represents the average
excess loss and coupling ratio for a single couple.
Each coupler has four measured excess losses and
coupling ratios, one pair for each (input) port.

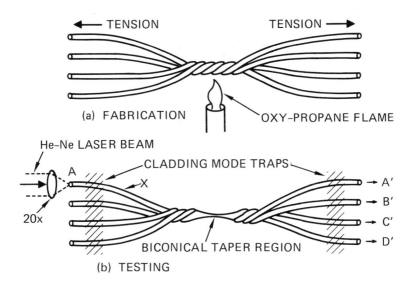

Figure 7.7   A four-channel bi-taper star coupler (after Ref. 7.4).

Fused biconical star couplers have also been fabricated using the same technique used to fabricate the T access couplers. The formation of a four-channel star coupler is illustrated in Figure 7.7. Transmissive stars with as many as 37 channels have been successfully fabricated (7.4). The excess insertion loss of these devices is approximately 2 dB. Transmissive star couplers have been used in a local computer network experiment called Fibernet. The Fibernet configuration used is shown in Figure 7.8. The "pseudo terminal" transmits pseudo-random data packets at 114 Mb/sec and checks for errors in the received packets. The demonstration employed a 19-channel transmissive star coupler as shown in the figure.

A fused biconical reflective star coupler has also been fabricated (7.5). A schematic for the design of an eight-port reflection-star coupler is shown in Figure 7.9. To fabricate such a device, four fibers of appropriate length are looped back on themselves, collected, and thermally fused. The measured coupling ratio for power coupled from port j to port i is listed in

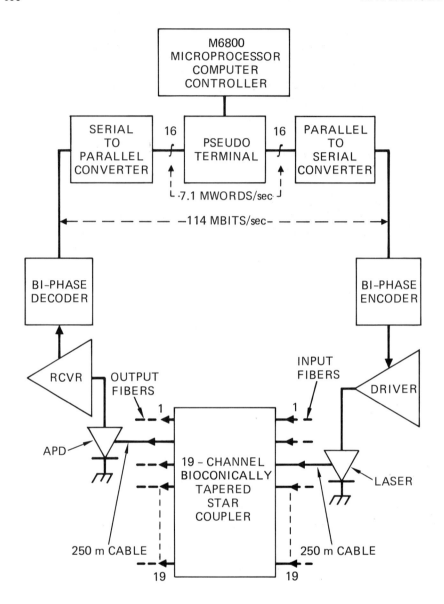

Figure 7.8. Fibernet configuration (after Ref. 7.4).

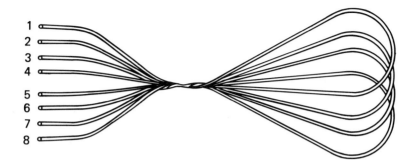

Figure 7.9.   A schematic of the design of an 8-port reflection-star coupler (after Ref. 7.5).

Table 7.1   Reflection-Star Coupler Performance (after Ref. 7.5)

Port Number	Input Power I_j, mW	Output Power P_i, mW	Coupling Ratio $100P_i/I_i$
1	4.85	(0.47)	9.7
2	—	0.5	10.3
3	—	0.525	10.8
4	—	0.42	8.7
5	—	0.855	17.6
6	—	0.43	8.9
7	—	0.44	9.1
8	—	0.53	10.9

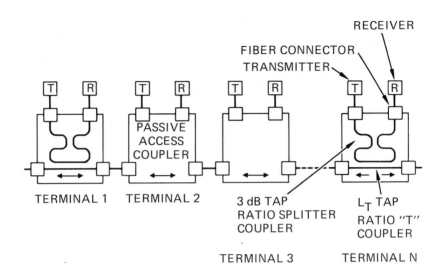

Figure 7.10.   Serial topology with passive access couplers employing fused fiber couplers.

Table 7.1.   A 0.7 dB excess insertion loss was determined.   In
Table 7.1, the output ports are labeled so that port i and port
(i+4) correspond to the same fiber.   As can been seen, the power
coupled from port 5 with port 1 excited is approximately twice that
of the other output powers.   Improved uniformity in the outputs
of this class of couplers can be achieved by cascading couplers
together as recently reported for transmission stars (7.6).
Coupler uniformity is, of course, important because it affects
the dynamic range requirements of the receiver.

7.5   DISTRIBUTION SYSTEMS

        Schematic diagrams of N terminal distribution systems
that were assembled using the fused-fiber couplers discussed in
the previous section are shown in Figures 7.10, 7.11, and 7.12.

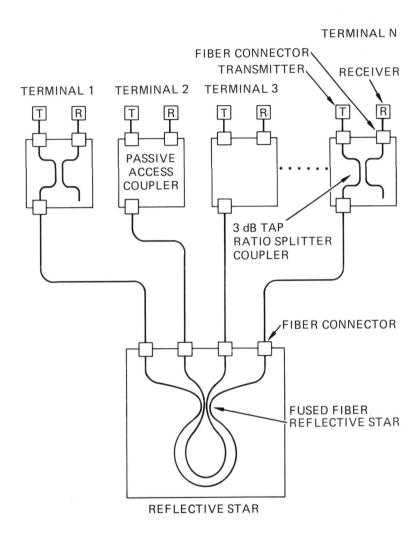

Figure 7.11. Reflection star topology employing fused fiber couplers.

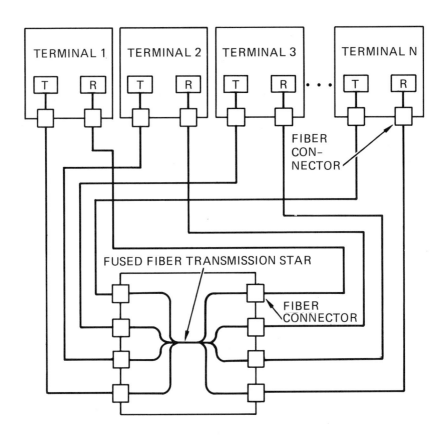

Figure 7.12. Transmission star topology employing fused fiber couplers.

The serial topology illustrated in Figure 7.10 utilizes a passive access coupler assembled from two fused T couplers. One T coupler provides bidirectional access to the data bus while the other serves as a 3 dB power divider. The worst case distribution system loss (excluding cable losses) for an N terminal system is given by

$$\frac{P_N(\text{received})}{P_1(\text{transmitted})} = 2(2L_c + L_s + L_T + L_{SCL} + L_{TCL})$$

$$+ (2L_c + L_{TCL} + L_{IT})(N-2) \quad ,$$

where

$$L_c \quad = \text{fiber connector loss}$$
$$L_s \quad = \text{splitting factor } (-3 \text{ dB})$$
$$L_T \quad = \text{T coupler tap ratio in dB}$$
$$L_{SCL} = \text{splitter coupler excess loss}$$
$$L_{TCL} = \text{T coupler excess loss}$$
$$L_{IT} \quad = \text{insertion loss associated with power}$$
$$\text{tapped by T coupler } (10 \log (1\text{-tap ratio})).$$

The star or parallel topology can be provided using either a reflection star coupler or a transmission star coupler. A parallel data network using a reflection star coupler is illustrated in Figure 7.11. Here the passive access coupler is comprised of a fused T power divider and the reflection star is of the fused fiber variety. The distribution system loss (excluding cable loss) for an N terminal system is given by

$$\frac{P_j(\text{received})}{P_i(\text{transmitted})} = 2(2L_c + L_{SCL} + L_s)$$

$$+ 2L_c + L_{RSL} + 10 \log 1/N \quad ,$$

where L_{RSL} is the excess loss of the reflection star coupler and
10 log 1/N is the power division factor, all other quantities
being previously defined.

The same network can be implemented by using the fused
fiber transmission star as illustrated in Figure 7.12. The distribution system loss for this case (again excluding cable) is
given by

$$\frac{P_j(\text{received})}{P_i(\text{transmitted})} = 2L_c + 2L_c + L_{TSL} + 10 \log 1/N \quad ,$$

where L_{TSL} is the excess insertion loss of the fused-fiber
transmission star.

Plots of distribution system losses for both the serial
and parallel formats as a function of the number of terminals are
shown in Figure 7.13. The parameters used for each case correspond to those that were demonstrated experimentally for fused
fiber couplers as discussed in the previous section. For simplicity, the experimentally observed port-to-port variation of the
output power of the star couplers has been neglected. The connector loss referred to in the figure is for the demountable
cable connectors required to attach both the access and star
couplers to the fiber-optic transmission line. A typical insertion loss for commercially available connectors is −0.75 dB. The
numerical values used for the serial access system illustrated in
Figure 7.10 are a 10 dB constant tap ratio T coupler with a
−0.5 dB excess insertion loss and a −3 dB tap ratio power
splitter with a −0.75 dB excess insertion loss. These values
were chosen from the data displayed in Figure 7.6. The coupler
parameters used for the reflection star system are a −3 dB tap
ratio power splitter with a −0.75 dB excess loss and a reflection
star coupler insertion loss of −0.7 dB, while a value of −2 dB
was used for the excess loss of the transmission star coupler.

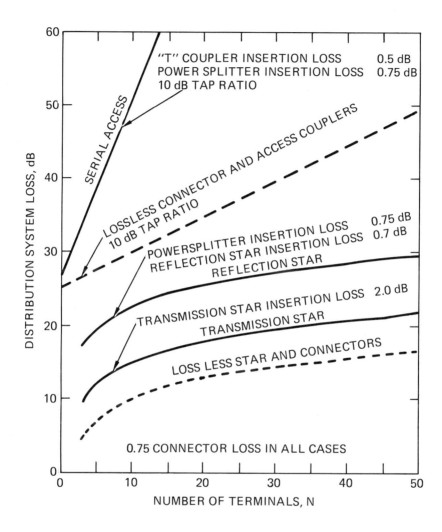

Figure 7.13. Distribution system loss plotted as a function of number of terminals.

Also shown in Figure 7.13 are the curves for both the parallel and serial systems that result if all the couplers and connectors are assumed lossless. The distribution system losses in this idealized case result only from splitting the power among the various terminals.

The comparison of the serial and parallel networks illustrates several features important in system design. The plots clearly reveal the signal level advantage of the star format over the serial format as the number of terminals in the system increases. The advantage is more pronounced as the insertion loss of both the cable connectors and the access couplers increase. However, it is relatively insensitive to the insertion loss of the star coupler. The greater distribution loss for the reflection star systems, compared with the transmission star system, results from the use of the passive access coupler to provide for power splitting between the transmitter and receiver. Note that the reflection star system requires only one half the amount of fiber length that the transmission star system needs.

For fixed-tap-ratio T couplers, the receiver in the serial system must be equipped with a wide dynamic range AGC to handle the strong signals from adjacent terminals and with the weak signals from remote terminals. Since the parallel system has but a single mixer, it does not have this dynamic range problem. The added uniform signal level available with the parallel system translates to less stringent design requirements on both the transmitters and the receivers. The cost of this added signal level is paid out in the amount of fiber cable necessary to wire the system. The star bus design in essence shortens the main bus to a single point mixer and extends the length of each terminal arm.

7.6   SYSTEM POWER BUDGET

To determine the amount of optical power available for
distribution and the maximum number of terminals that can be
served by the serial and star topologies, assume that the maximum
distance between the first and last terminal in the N terminal
serial system is 1 km.   That is, the sum of all terminal-to-
terminal spacings does not exceed 1 km.   Assuming a 5 dB/km fiber
cable with a 0.2 NA fiber, the amount of optical power coupled
into the cable at the input would be approximately 0 dBm for an
injection laser diode (ILD) source and −9 dBm for a Burrus type
LED.   For data transferred at the rate of 50 Mb/sec with a bit
error rate of 10^{-9}, the power required at the receiver is approximately −42 dBm for a PIN receiver and −57 dBm for an optimized
APD receiver.   For the 1 km serial system the distribution loss
power budget is therefore −28 dB for the LED/PIN combination and
−52 dB for the ILD/APD combination.   Reference to Figure 7.13
reveals that the maximum number of terminals that can be accessed
without repeaters is zero and ten, respectively.   (The bit error
rate could not be maintained if the LED/PIN combination were
used.)   If perfect connectors and couplers were used, the maximum
terminal numbers are 6 and 55.   On the other hand, even if each
terminal in either the transmission or reflection star system was
located 1/2 km from the star, the number of terminals that could
be included in the network is far in excess of 50 using either
source/receiver combinations.

7.7   SUMMARY

In this chapter an elementary treatment of some of the
engineering design considerations that must be addressed in designing data distribution systems using fiber-optic transmission

lines has been presented. Serial and parallel distribution system assemblies, using passive coupling components that have been experimentally demonstrated, have been compared.

PROBLEMS FOR CHAPTER 7

1.   For a 100 Mb/sec data rate and a bit error rate of 10^{-9}, how much optical power is required incident on an APD receiver with optimum gain? What is the corresponding number for a PIN receiver?

2.   Referring to the optical power-versus-current curves in Figure 7.3, discuss the advantages and disadvantages of using an ILD and LED as a light source.

3.   Describe qualitatively the principle of operation of the fused fiber coupler.

4.   Assume a collection of terminals distributed over a geographic area defined by a circle of 1 km diameter. (Physically distribute the terminals as desired.) For this network of terminals, how many terminals can be served in a repeaterless serial system, reflection star system, and transmission star system?

5.   For the system defined in Problem 4, how much gain would a repeater require to double the amount of terminals accessible with the serial system?

6.   What are the distribution system losses for the reflection and transmission star systems if the number of terminals to be served is 1,000?

7.   For the network defined in Problem 4, how much fiber was employed for the serial, reflection star, and transmission star systems?

8.   Assuming uniform star couplers, determine the maximum variation in received power between the terminals for the network defined in Problem 4. Which of the three systems do you consider best?

REFERENCES

7.1 Figure 7.2 is identical to Figure 6.14. It is reproduced
     here for convenient reference.

7.2 M.K. Barnoski and H.R. Friedrich, *Applied Optics 15*, 2629
     (1976).

7.3 B.S. Kawasaki and K.O. Hill, *Applied Optics 16*, 1795 (1977).

7.4 E.G. Rawson, *Optical Fiber Communication Technical Digest*,
     IEEE Catalog Number 79CH1431-6 QEA, p. 60 (1979).

7.5 D.C. Johnson, B.S. Kawasaki, and K.O. Hill, *Applied Physics
     Letters 35*, 479 (1979).

7.6 A. Yoshida, R.G. Lamont, and D.C. Johnson, "Cascaded Trans-
     mission Star Couplers," to be presented at the 1981 CLEO
     Conference, Washington, D.C., June 10-12, 1981.